The Recursive Mind

The Origins of Human Language,
Thought, and Civilization

Michael C. Corballis

The Recursive Mind

The Recursive Mind

The Origins of Human Language, Thought, and Civilization

Michael C. Corballis

PRINCETON UNIVERSITY PRESS   PRINCETON AND OXFORD

Published by Princeton University Press, 41 William Street,
Princeton, New Jersey 08540
In the United Kingdom: Princeton University Press, 6 Oxford Street,
Woodstock, Oxfordshire OX20 1TW
press.princeton.edu

Library of Congress Cataloging-in-Publication Data

Corballis, Michael C.
 The recursive mind : the origins of human language, thought, and civilization /
Michael C. Corballis.
     p. cm.
 Includes bibliographical references and index.
 ISBN 978-0-691-14547-1 (cloth : alk. paper)
1. Evolutionary psychology. 2. Human evolution.
3. Brain—Evolution. 4. Language and languages—Origin.   I. Title.
 BF701.C665 2011
 155.7—dc22   2010043084

British Library Cataloging-in-Publication Data is available

This book has been composed in Sabon LT Std text with Scala Sans OT Condensed display
Printed on acid-free paper ∞
Printed in the United States of America
10 9 8 7 6 5 4 3 2 1

Contents

Preface

We humans like to think that we have capacities that make us not only distinct from all other creatures on the planet, but also superior to them. What other species, you might ask, has measured the speed of light, figured out how the universe began, invented the laptop computer, or painted a portrait? Our species has even succeeded in escaping from the planet altogether, even if only fleetingly. You might also ask, I suppose, why any other species would care to do these things, and we do need to be wary of our comfortable assumption that we are at the top of the earthly hierarchy, since it provides a too-easy justification of the appalling way we treat other animals. Let's face it: We eat them, kill them for sport, drink their milk, wear their skins, ride on their backs, ridicule them, house them in zoos, and breed them to our own specifications.

By the same token, though, it cannot be denied that our species has dominated the earth like no other. Not only do we subjugate other creatures to our needs and whims, but we also mold the physical environment to our specifications, to the point that our success may prove to be our undoing. Unless we make better use of our vaunted intelligence, we run the risk of succumbing to pollution, global warming, or weapons of mass destruction—or, to think recursively, of weapons for the mass destruction of weapons of mass destruction. And yet we are biologically almost indistinguishable from the other great apes, and share a common ancestor with the chimpanzee and bonobo dating from only about six or seven million years ago—a mere eye-blink in evolutionary time. In marked contrast to human triumphalism, the great apes have been forced into ever-diminishing habitats, and they too are threatened with extinction.

Many have conjectured about why our species is so dominant on the planet. Assuredly, the reason is mental rather than physical—any number of animals out there can easily beat us in physical combat. Descartes argued that only humans are capable of free will. Aristotle suggested that man is the only political animal, and history suggests that he should have included women. Thomas Willis thought that only humans were capable of laughter, while Martin Luther argued that it was the possession of property that distinguished us. Benjamin Franklin attributed human uniqueness to tool-making, and the Greek philosopher Anaxagoras said it was the human hand that made us the wisest of species. Steven Mithen recently suggested that music may have started it all. Some years ago, in my book *The Lopsided Ape*, I argued that the asymmetry of the human brain was what made us what we are. There is probably some truth to at least some of these assertions, but readers may observe that lopsidedness receives very little mention in this book.

The one characteristic that has received most attention is language. "In the beginning," said St. John, "was the Word, and the Word was with God, and the Word was God" (John 1:1). In the seventeenth century, Réné Descartes argued that language, as an expression of free will, was so unconstrained that it could not be explained in terms of mechanical principles, and must therefore have been a gift from God. In the following century, another French philosopher, Abbé Étienne Bonnot de Condillac, speculated about how language evolved, but as an ordained priest he was fearful of offending the church, and disguised his theory in the form of a fable—as we shall see in chapter 4. In 1866, the Linguistic Society of Paris banned all discussion of the origins of language.

In the twentieth century the linguist Noam Chomsky, himself a self-styled Cartesian, also argued that language could not have evolved through natural selection. His reasoning was based not on religion, but rather on a view of how language works. Basically, he argued that external language, as spoken or signed, must have arisen from an internal language—essentially the "language of thought"—that has no direct reference to the outside world, and so could not have been subjected to the pressures of adaptation to the environment. Chomsky therefore argued that internal language emerged from some single and singular event causing a rewiring, perhaps a fortuitous mutation, of the brain. He argued further that this event occurred late in the evolution of our species, perhaps even within the past 100,000 years. This account, although not driven by religious doctrine, does smack of the miraculous.

Chomsky is nevertheless one of the heroes of this book. He has long recognized the open-ended nature of language, and suggested that the key to this open-endedness is recursion. By applying principles in recursive fashion, we can create utterances, whether spoken or signed, of essentially infinite variability. Where I part from Chomsky, though, is in his view that thought itself is fundamentally linguistic. I argue instead that the modes of thought that made language possible were nonlinguistic, but were nonetheless possessed of recursive properties to which language adapted. Where Chomsky views thought through the lens of language, I prefer to view language through the lens of thought. This change of view provides the main stimulus for this book, since it not only leads to a better understanding of how we humans think, but it also leads to a radically different perspective on language itself, as well as on how it evolved.

I focus on two modes of thought that are recursive, and probably distinctively human. One is mental time travel, the ability to call past episodes to mind and also to imagine future episodes. This can be a recursive operation in that imagined episodes can be inserted into present consciousness, and imagined episodes can even be inserted into other imagined episodes. Mental time travel also blends into fiction, whereby we imagine events that have never occurred, or are not necessarily planned for the future. Imagined events can have all of the complexity and variability of language itself. Indeed I suggest that language emerged precisely to convey this complexity, so that we can share our memories, plans, and fictions.

The second aspect of thought is what has been called theory of mind, the ability to understand what is going on in the minds of others. This too is recursive. I may know not only what you are thinking, but I may also know that you know what I am thinking. As we shall see, most language, at least in the form of conversation, is utterly dependent on this capacity. No conversation is possible unless the participants share a common mind-set. Indeed, most conversation is fairly minimal, since the thread of conversation is largely assumed. I heard a student coming out of a lecture saying to her friend, "That was really cool." She assumed, probably rightly, that her friend knew exactly what "that" was, and what she meant by "cool."

That, then, is the theme of the book, but there are many excursions—into such questions as whether animals have language, whether human language evolved from manual gestures, whether all languages share common principles, why fiction is adaptive. Given the view that human thought and language evolved gradually, I sketch the likely evolution of our distinctive characteristics over the past 6 million or so years, and not, as Chomsky would have it, over the past 100,000 years. And if you don't understand what recursion is, I hope this book will give you a better idea.

Many people have inspired my thinking scientifically and philosophically, and of course many of them, perhaps most of them, will disagree with at least some of my conclusions. They include Donna Rose Addis, John Andreae, Michael Arbib, Giovanni Berlucchi, Brian Boyd, Noam Chomsky, Nicola Clayton, Erica Cosentino, Karen Emmorey, Nicholas Evans, Francesco Ferretti, Tecumseh Fitch, Maurizio Gentilucci, Russell Gray, Nicholas Humphrey, Jim Hurford, Steven Pinker, Giacomo Rizzolatti, Michael Studdert-Kennedy, Thomas Suddendorf, Endel Tulving, and Faraneh Vargha-Khadem. I especially thank my wife, Barbara, for putting up with the hours I spent over the computer; at least she had her golf. My sons Tim and Paul—the latter claimed to a friend that he taught me everything I know—have often corrected me on points of psychology and philosophy.

I also owe a special debt of thanks to Eric Schwartz, Beth Clevenger, Richard Isomaki, and Jeffrey Weiss of Princeton University Press, and to my agent Peter Tallack, for their invaluable help in putting some shape into the book.

The Recursive Mind

1

What Is Recursion?

In 1637, the French philosopher Réné Descartes wrote the immortal line "Je pense, donc je suis." Curiously, this is usually rendered in Latin, as *Cogito, ergo sum*, and is translated in English as "I think, therefore I am." In making this statement, Descartes was not merely thinking, he was thinking about thinking, which led him to the conclusion that he existed. The recursive nature of Descartes's insight is perhaps better rendered in the version offered by Ambrose Bierce in *The Devil's Dictionary*: *Cogito cogito ergo cogito sum*—"I think I think, therefore I think I am." Descartes himself, though, was more prone to doubt, and expanded his dictum as "Je doute, donc je pense, donc je suis"—"I doubt, therefore I think, therefore I am." He thus concluded that even if he doubted, someone or something must be doing the doubting, so the very fact that he doubted proved his existence. This probably came as a relief to his friends.

In this book, I examine the more general role of recursion in our mental lives, and argue that it is the primary characteristic that distinguishes the human mind from that of other animals. It underlies our ability not only to reflect upon our own minds, but also to simulate the minds of others. It allows us to travel mentally in time, inserting consciousness of the past or future into present consciousness. Recursion is also the main ingredient distinguishing human language from all other forms of animal communication.

Recursion, though, is a fairly elusive concept, often used in slightly different ways.[1] Before I delve into some of the complexities, let's consider some further examples to give the general idea. First, then, a not-too-serious dictionary definition:

Recursion (rĭ-kûr'-zhən) *noun*. See **recursion**.

One problem here, of course, is that this implies an infinite loop,

Figure 1. The thinker thinks of thinking of thinking (author's drawing).

from which you may never escape in order to read the other stuff in this book. The following variant suggests a way out:

> **Recursion** (rĭ-kûr'-zhən) *noun*. If you still don't get it, see **recursion**.

This banks on the possibility that if you do get it after a round or two, you can escape and move on. If you don't, well I'm sorry.

The postmodern novelist John Barth concocted what is probably both the shortest and the longest story ever written, called *Frame-Tale*. It can be reproduced as follows: Write the sentence *ONCE UPON A TIME THERE* on one side of a strip of paper, and *WAS A STORY THAT BEGAN* on the other side. Then twist one end once and attach it to the other end, to form a Mobius strip. As you work your way round the strip, the story goes on forever.

A similar example comes from an anonymous parody of the first line of Bulwer-Lytton's infamous novel, *Paul Clifford*:

> It was a dark and stormy night, and we said to the captain, "Tell us a story!" And this is the story the captain told: "It was a dark and stormy night, and we said to the captain, 'Tell us a story!' And this is the story the captain told: 'It was a dark . . .' "

Another amusing example is provided by a competition, run by *The Spectator* magazine, which asked readers to state what they would most like to read on opening the morning paper. The winning entry read as follows:

Our Second Competition

The First Prize in the second of this year's competitions goes to Mr Arthur Robinson, whose witty entry was easily the best of those we received. His choice of what he would like to read when opening the paper was headed, "Our Second Competition," and was as follows: "The First prize in the second of this year's competitions goes to Mr Arthur Robinson, whose witty entry was easily the best of those we received. His choice of what he would like to read when opening the paper was headed 'Our Second Competition,' but owing to paper restrictions we cannot print all of it."[2]

Taking a different tack, John Barth's story *Autobiography: A Self-recorded Fiction* is a recursive tale in which the narrator is ostensibly the story itself, writing about itself.[3] It ends, recursively, in its own end:

Nonsense, I'll mutter to the end, one word after another, string the rascals out, mad or not, heard or not, my last words will be my last words.

To my knowledge, no story has yet attempted to write a story of a story that writes about itself.

And then there is the recurring problem of fleas, as penned by the Victorian mathematician Augustus de Morgan:

Great fleas have little fleas upon their backs to bite 'em,
And little fleas have smaller fleas, and so *ad infinitum*.
And the great fleas themselves, in turn, have greater fleas to go on,
While these again have greater still, and greater still, and so on.[4]

This notion of inserting progressively smaller entities into larger ones *ad infinitum* can also give rise to interesting visual effects, as in the examples shown in figure 2.

Apollonian gasket   Sierpinski triangle

Figure 2. Two figures showing recursive insertions of circles (*left*) and triangles (*right*). The Apollonian gasket derives from Apollonius, a Greek geometer from the third century BC, who studied the problem of how to draw a circle that is tangential to three circles. Starting with three circles that are tangential to one another, one can continue the process of constructing circles tangential to all triplets *ad infinitum*. The resulting figure serves as a mathematical model for foam (see Mackenzie 2009 for more information). (2D Apollonian gasket with four initial circles courtesy of Guillaume Jacquenot.)

The use of recursion to create infinite sequences is also exploited by mathematics. One such sequence is the set of natural (i.e. whole) numbers, which I'll write as **N**. Thus we can generate all of the positive natural numbers by the definitions

1 is in **N**

If n is in **N** then (n + 1) is in **N**.

This second definition is recursive, because **N** appears in the condition that needs to be satisfied for **N**.

You may remember, possibly from schooldays, what *factorials* are. As a schoolboy I found them amusing in a childish kind of way, because they were signaled with exclamation marks; thus factorial 3, usually written 3!, is 3 * 2 * 1, and equals 6.[5] Similarly, we can compute the following:

$$5! = 5 * 4 * 3 * 2 * 1 = 120$$
$$8! = 8 * 7 * 6 * 5 * 4 * 3 * 2 * 1 = 40,320$$

Clearly, this can go on forever, but we can capture the entire set by using just two defining equations:

$$0! = 1$$
$$n! = n * (n - 1)! \quad [\text{where } n > 0].$$

This second equation is recursive in that a factorial is defined in terms of a factorial. We need the first equation to kick the thing off.

The next example is for rabbits, and is called the Fibonacci series, defined by the following three equations:

fibonacci(0) = 1
fibonacci(1) = 1
fibonacci(n) = fibonacci(n − 1) + fibonacci (n − 2)
[where $n > 1$].

If you are following me, you should be able to compute the series, which goes 1, 1, 2, 3, 5, 8, 13, ... What the definition says, then, is that each number in the series is the sum of the two previous ones. Why rabbits? Fibonacci (c. 1170–1250) was an Italian mathematician who used the series to predict the growth of a hypothetical population of rabbits.[6]

For a final informal example, I take you to Kyoto, Japan, where I once happened upon a sign on a gate that was written in Kanji script. I asked what it meant, and my guide told me, I hope correctly, that it meant *Post no bills*. There is a paradox here in that the sign was itself a bill, thereby contravening its own presence. Perhaps there needed to be another sign that said *Post no "Post no bills" bills*. But of course this is itself in violation of its own message, so we might envisage another sign that reads *Post no 'Post "Post no bills" bills' bills*. There is no end to this process, so it might have been more sensible to have allowed bills on the gate in the first place. In practice, though, limitations of time, space, or memory will prevent a recursive sequence of structure continuing forever.

Toward a Working Definition

One of the characteristics of recursion, then, is that it can takes its own output as the next input, a loop that can be extended

autopoeisis ?

indefinitely to create sequences or structures of unbounded length or complexity. In practice, of course, we do not get caught up in infinite loops—life is simply too short for that. For the purposes of this book, then, we shall not be interested so much in the generation of infinite sequences as in a definition that might apply usefully to human thought. A definition that meets this requirement is suggested by Steven Pinker and Ray Jackendoff, who define recursion as "a procedure that calls itself, or . . . a constituent that contains a constituent of the same kind."[7]

The second part of this definition is important, especially in language, because it allows that recursive constructions need not involve the embedding of the *same* constituents, as in the example of the gate in Kyoto, but may contain constituents of the same *kind*—a process sometimes known as "self-similar embedding." For example, noun phrases can be built from noun phrases in recursive fashion. Tecumseh Fitch gives the example of simple noun phrases such as *the dog, the cat, the tree, the lake,* and one can then create new noun phrases by placing the word *beside* between any pair: *the dog beside the tree, the cat beside the lake.*[8] Or one might have two sentences: *Jane loves John* and *Jane flies airplanes,* and embed one in the other (with appropriate modification) as *Jane, who flies airplanes, loves John.* These can be extended recursively to whatever level of complexity is desired. For example we could extend the noun phrase to *the dog beside the tree beside the lake,* or the sentence about Jane and John to *Jane who flies airplanes that exceed the sound barrier loves John, who is prone to self-doubt.* Most languages make use of recursive operations of this sort—although we shall see in the next chapter that there may be a few languages that don't operate in this way.

Although it is common to provide illustrations from language, the main theme of this book is that it is in thought rather than in language that recursion originates. As Pinker and Jackendoff put it, "The only reason language needs to be recursive is because its function is to express recursive thoughts. If there were not any recursive thoughts, the means of expression would not need recursion either."[9] In remembering episodes from the past, for instance, we essentially insert sequences of past consciousness into present consciousness, or in our interactions with other people we may insert what they are thinking into our own thinking. These themes are explored in later chapters.

Process and Structure

As suggested by Pinker and Jackendoff's dual definition, recursion can be understood either as a *process* or as a *structure*. The distinction can be important. A recursive process may lead to a structure that need not be seen as itself recursive. For example, suppose we construct a sequence of musical notes with an embedding routine by pairing pairs of notes, each consisting of a randomly chosen note played on a piano with a randomly chosen note played on a violin. The first pair is embedded in another pair, and the four-note output then embedded in another pair. This process can be continued indefinitely to create a sequence of notes. As illustrated in figure 3, though, the sequence can be interpreted, not as a recursively embedded structure, but as a sequence of piano notes followed by an equally long sequence of violin notes. This failure to distinguish recursive embedding from recursive structure has led to some confusion, especially in claims about recursion in nonhuman species.[10]

Again, in his most recent theory on the nature of language, known as the Minimalist Program,[11] Noam Chomsky has argued that human thought is generated by a Merge operation, applied recursively. That is, units are merged to form larger entities, and the merged entities can be themselves merged to form still larger entities, and so on. This operation underlies the embedded structure of human language, although in Chomsky's theory it applies strictly to what he calls *I-language*, which is the thought process preceding *E-language*, the external language that is actually spoken or signed. Merge can produce strings of elements, be they words or elements of thought, and although it may be applied recursively to produce hierarchical structure, that structure may not be evident in the final output. For instance, even sentences might be regarded simply as words all merged in unstructured sequence, as in ritualized songs or prayers. Everyday language, too, may include mentally undifferentiated clichés and slogans, or sequences that are

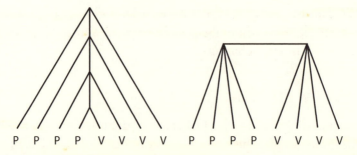

P P P P V V V V   P P P P V V V V

Figure 3. The sequence of Ps and Vs can be created either by recursively nesting PV pairs in PV pairs (*left*), or by arranging a sequence of Ps followed by a sequence of an equal number of Vs (*right*). The sequence might be generated as in the left panel, but interpreted as in the right panel.

highly automated. Politicians may be especially prone to this kind of talk.

As noted above, recursive processes and structures can in principle extend without limit, but are limited in practice. Nevertheless recursion does give rise to the *concept* of infinity, itself perhaps limited to the human imagination. After all, only humans have acquired the ability to count indefinitely, and to understand the nature of infinite series, whereas other species can at best merely estimate quantity, and are accurate only up to some small finite number.[12] Even in language, we understand that a sentence can in principle be extended indefinitely, even though in practice it cannot be—although the novelist Henry James had a damn good try. Such understandings are indeed part of human mental achievement, and depend on a human capacity for recursive thought. Nevertheless they are not the primary concerns of this book.

The appealing aspect of recursion is precisely that it can *in principle* extend indefinitely to create thoughts (and sentences) of whatever complexity is required. The idea has an elegant simplicity, giving rise to what Chomsky called "discrete infinity,"[13] or Wilhelm Humboldt (1767–1835) famously called "the infinite use of finite means." And although recursion is limited in practice, we can nevertheless achieve considerable depths of recursive thought, arguably unsurpassed in any other species. In chess, for example, a player may be able to think recursively three or four steps ahead, examining possible moves and countermoves, but the number of possibilities soon multiplies beyond the capacity of the mind to hold them.

Deeper levels of recursion may be possible with the aid of writing, or simply with extended time for rehearsal and contemplation, or extended memory capacity through artificial means. The slow development of a complex mathematical proof, for example, may require subtheorems within subtheorems. Plays or novels may involve recursive loops that build slowly—in Shakespeare's *Twelfth Night*, for example, Maria foresees that Sir Toby will eagerly anticipate that Olivia will judge Malvolio absurdly impertinent to suppose that she wishes him to regard himself as her preferred suitor.[14] As in fiction, so in life; we all live in a web of complex recursive relationships, and planning a dinner party may need careful attention to who thinks what of whom.

The structures resulting from recursive processes need not reveal the nature of those processes, just as a loaf of bread may not reveal the processes of kneading that went into the making of the bread, or the taste of wine the picking and trampling of the grapes. Often, though, the structure of a sentence or stream of thought may reveal recursive embedding—interpretation of a sentence may require the understanding of phrases embedded in phrases, regardless of how the embedding was actually accomplished, and an internal understanding of a stream of thought may require the segmentation of episodes within episodes.

What Recursion Is Not

Recursion is not the only device for creating sequences or structures of potentially infinite length or size. I now consider some examples that do not meet the criteria for recursion.

Repetition

Simple repetition can lead to sequences of potentially infinite length, but does not classify as true recursion. For example, the sentence that opens chapter 9 of A. A. Milne's *Winnie the Pooh* goes *It rained and it rained and it rained.* This could go on for-ever—or at least until Piglet is drowned—but the repetition simply conveys the information that it rained rather a lot, causing Pig-let some ennui. It is not recursive because each addition of *and it rained* is not driven by the previous one; it is simply added at the discretion of the writer.

In any event, repetition does not distinguish human activity from that of nonhuman animals. Birdsong, for example, is relentlessly repetitive, but each repeated theme does not embellish or qualify the previous one. At most, the repetition might signal urgency, or simply signal continuing presence, as one might repeatedly knock on a door in the hope of arousing someone inside. Repetition is ubiquitous in human and animal life, in activities ranging from the repeated jaw movements in eating, to the curiously repetitive na-ture of sexual activity. The spider, no less, is capable of repetition, as in Walt Whitman's *Leaves of Grass*:[15]

> A NOISELESS, patient spider
> I mark'd, where, on a little promontory, it stood, isolated;
> Mark'd how, to explore the vacant, vast surrounding,
> It launch'd forth filament, filament, filament, out of itself;
> Ever unreeling them—ever tirelessly speeding them

Information can also be aggregated in nonrecursive fashion, as when the short-story writer Saki (H. H. Munro) wrote, "Hunger, fatigue, and despairing hopelessness had numbed his brain."[16] Ag-gregation of different phrases similarly compounds meaning ad-ditively, as when the historian Peter Hennessy wrote:

> The model of a modern Prime Minister would be a kind of grotesque composite freak—someone with the dedication to duty of a Peel, the physical energy of a Gladstone, the detachment of a Salisbury, the balls of a Lloyd George, the word-power of a Churchill, the administrative gifts of an Attlee, the style of a Macmillan, the managerialism of a Heath, and the sleep requirements of a Thatcher.[17]

The sentence itself has recursive elements, but the aggregation of phrases to describe the freakish composite is not recursive in that each does not call the next. Instead, they are effectively elements in a list, inserted to add information. Nonhuman species may well have a similar ability to accumulate information, as when understanding a predator as large, fierce, and with sharp teeth and claws.

Iteration

A slightly more subtle variant on repetition and aggregation is *iteration*, where a process is repeated, but in this case there is input from the previous application of the process. In this respect it is like recursion, and indeed considered by mathematicians to belong to the class of "general recursive functions." For the main purposes of this book, though, it does not qualify as true recursion because each output is discarded once it has been entered into the next application. The dictionary definition of recursion that I gave earlier in this chapter was also really an example of iteration rather than recursion, because you just keep going round and round the loop, without any added structure. The iterations therefore do not lead to added complexity.[18]

Iterative procedures are used in computational mathematics to arrive at increasingly accurate solutions to a problem. The basic idea is to start with a preliminary solution—perhaps a guess—and then use a procedure to compute a new solution. This solution is then used as the starting point for the next computation, and the new solution is then the starting point for the next round. The cycle is repeated until the solutions stabilize to some acceptable criterion.[19] Feedback systems operate in much the same way, typically as a means of maintaining homeostasis. For example, a thermostat may involve a system for raising or lowering temperature, and the goal is to achieve some given temperature. The actual temperature is fed into the system, which operates to raise or lower the temperature until the desired is reached. The body is awash with feedback systems to maintain homeostasis of temperature, iron, energy, blood composition, and so on. The main regulator is the hypothalamus, in the limbic system of the brain. Such systems again do not differentiate humans from other animals.

Sometimes the distinction between recursion and iteration may be a matter of interpretation. In the infinite loop created by the parody of *Paul Clifford*, one might say that each beginning of the story is initiated by the previous one, which is then forgotten. The parody is best appreciated, though, if the story is seen as an endless, ever deepening whirlpool, with each segment remaining as part of it. I'm told the story works best if each segment is spoken with a different accent.

Consider too this line from a well-known children's verse:

> This is the dog that worried the cat that killed the rat that ate the malt that lay in the house that Jack built.

To understand this sentence as truly recursive, one must appreciate that it describes a state of affairs as a complex whole, and refers to particular cases of a dog, cat, rat, malt, house, and fellow called Jack. It is not simply the stringing together of a dog that worried a cat, a cat that killed a rat, and so forth. A young child, though, might process it in this piecemeal way, as a succession of unrelated events.

Recursion and Evolutionary Psychology

In emphasizing recursion as a unifying concept, the approach taken in this book contrasts with that adopted by so-called evolutionary psychologists, who have argued that the mind has multiple facets. The basic tenets of evolutionary psychology were laid out in the 1992 volume *The Adapted Mind*, edited by Jerome Barkow, Leda Cosmides, and John Tooby, and popularized by Steven Pinker in his influential 1997 book *How the Mind Works*.[20] Thus Pinker writes that the human mind "is not a single organ but a system of organs, which we can think of as psychological faculties or mental modules."[21] In examining present-day human behavior, the evolutionary psychologist's agenda is to discover independent processes as the basic modules, and relate them to conditions that prevailed in the Pleistocene, when humans existed primarily as hunter-gatherers. As Pinker puts it, the aim is to carve the mind at its joints, so to speak, and "reverse-engineer" its components, or modules, back to the epoch during which the human mind was formed.

In this view, the mind is really a collection of miniminds, each beavering away on its own specific problem, among which are language and theory of mind. This has been called the Swiss-army-knife model of the mind, with a blade for every purpose.[22]

The danger with this approach is that it becomes too easy to postulate modules and to tell "just so" stories about how they evolved, so that there is a risk of returning to the now-abandoned instinct psychology of the early twentieth century.[23] Instinct psychology perished under the sheer weight of numbers—the author of one text counted 1,594 instincts that had been attributed to animals and humans[24]—and evolutionary psychology may also drown in a sea of modules, if not of mixed metaphors. Pinker suggests that we like potato chips because fatty foods were nutritionally valuable during the Pleistocene, but scarce enough that there was no danger of obesity; we like landscapes with trees because trees provides shade and escape from dangerous carnivores on the Africa savanna; flowers please us because they are markers for edible fruits, nuts, or tubers amid the greenery of the savanna; and so on. "There are modules," he writes, "for objects and forces, for animate beings, for minds, and for natural kinds like animals, plants, and minerals."[25]

This is not to say that the Swiss-army-knife model is without merit. Some of the postulated modules do provide insight into the human condition, and are reasonably well founded. For example, foundational work in evolutionary psychology by Leda Cosmides provided good evidence for a "cheater-detection module"—an instinctive ability to detect those who flout social conventions for their own gain.[26] A recent study suggests that humans possess a "category-specific attention system" that is especially adapted for attending to animals;[27] one of the authors, John Tooby, is quoted as saying, "Even dull animals like pigeons . . . recruit a surprising amount of attention—as do turtles resembling rocks."[28] This book is not intended to deny that there are many specific dispositions that shape our mental and social lives; rather, my aim is to suggest that there are deeper aspects of human thought that are governed by similar principles, and that recursion is one of those principles—and perhaps the most important one.

To be fair, too, not all evolutionary psychologists have insisted that modules are completely encapsulated, shut off from any communication with one another. Even Steven Pinker, for example, writes, "[Modules] accomplish specialized functions, thanks to their specialized structures, but don't necessarily come in encapsulated packages."[29] Steven Mithen, although scarcely a card-carrying evolutionary psychologist, argued that the human mind evolved its distinctive character precisely because previously encapsulated modules began to "leak," creating what he calls "cognitive fluidity."[30] It is as though the modules stopped minding their own businesses, and began to gossip. My approach in this book is not entirely at odds with this view, in that I argue a common principle might underlie a number of our distinctive abilities.

Others are also beginning to question the Swiss-army-knife model of the human mind more starkly. David Premack, for example, adopts an approach similar to that offered in this book. Reviewing the evidence for discontinuity between humans and other animals, he writes: "Animal competencies are mainly adaptations restricted to a single goal. Human competencies are domain-general and serve numerous goals."[31] This in effect reverses the evolutionary psychology argument—the mind has become less rather than more modular. The tide may well be turning.

In any event it is unlikely that recursion can be considered a module. As we shall see, recursion seems to be an organizing principle in very different spheres of human mental activity, from language to memory to mind reading. Recursive thinking probably depends on other mental attributes. One of these is what has been termed working memory, which holds information in consciousness. In order to embed processes within processes it is necessary to remember where one had got to in the earlier process when an embedded process has been completed. For instance, in a sentence like *My dog, who eats bananas, often gets sick,* one must hold the early part of the sentence (*My dog*) and link it to the next part (*often gets sick*). Dwight W. Read has argued that nonhuman primates, even our closest relatives the chimpanzees, have a working memory that is too limited to allow this kind of embedding.[32] Recursion probably also depends on an executive process that organizes what is to be embedded in what, and this may depend on the frontal lobes of the brain. The ability to organize and carry out recursive operations may therefore depend on several processes.

Although I do not embrace the modular view assumed by evolutionary psychologists, I am at one with them in proposing that a distinctively human mind evolved during the Pleistocene, the epoch that stretched from around 2.6 million years ago to some 12,000 years ago. How that happened will be told in the later chapters of this book.

Plan of the Book

The book is divided into four parts.

Part 1 deals with language. Although recursion is not limited to language, it is most commonly invoked to explain why human language differs from other forms of animal communication, an insight largely attributable to Noam Chomsky. Chapter 2 discusses the nature of language, with particular emphasis on the role of recursion. Chapter 3 then raises the age-old question of whether other animals have anything resembling human language. Chapter 4 develops the idea that language evolved from manual gestures— an idea that suggests greater evolutionary continuity between humans and other primates than the more common assumption that language emerged from vocal calls.

Part 2 deals with mental time travel, the ability to bring to mind events removed from the present in both time and place. Chapter 5 starts with memory, and develops the idea that memory for specific episodes is unique to humans. Chapter 6 extends the notion of episodic memory to the imagining of possible future events, leading to the concept of the self as existing through time. This leads to the notion, discussed in chapter 7, that language itself evolved to enable people to share their memories and plans, and so to communicate about events that are not present in the immediate environment. This leads also to fiction—the telling of stories that need not be based on fact, but that nonetheless hone the capacity to deal with the episodic exigencies of human social life.

Part 3 deals another recursive aspect of human thought, namely, theory of mind—or the ability to understand what others are thinking or feeling. Chapter 8 introduces mind-reading, not as a psychic phenomenon, but as a natural ability to infer the mental perspectives of other people. This ability is again critical to social cohesion and cooperation. Chapter 9 explains how theory of mind was also critical to the emergence of language.

Part 4 delves more specifically into the question of how the recursive mind evolved. Chapter 10 sets this question in the context of the classic debate between Cartesian discontinuity and Darwinian continuity. Chapter 11 examines some of the steps by which the hominins,[33] after splitting from the line leading to modern chimpanzees and bonobos, began to assume human-like attributes. Chapter 12 then considers the final step to "modern" *Homo sapiens*, the sole surviving hominin species—dominant, manipulative, Machiavellian, and capable of pondering our own nature and status on the planet. That, perhaps, is the ultimate triumph of the recursive mind.

Chapter 13 presents the final summary and conclusions.

PART 1

Language

Language takes pride of place because it is often considered the one faculty that makes humans unique, although we shall consider other candidates in later chapters. In the Chomskyan view of language, moreover, recursion is seen as its most distinguishing feature. As we shall see in the following three chapters, though, this view is undergoing some revision, and there may even be some languages that do not make use of recursive principles. Moreover, closer scrutiny of animal communication suggests greater continuity than a strict Chomskyan or Cartesian perspective might imply. The notion of continuity is buttressed by the argument that language evolved from manual gestures rather than from vocal calls, as I explain in chapter 3.

Even so, language remains vastly more complex than any form of animal communication, and understanding how it evolved, it has been suggested, might be "the hardest problem in science."[1] Language may not always draw on recursive principles, but a main theme of this book is that it nevertheless depends on the recursive nature of nonlinguistic thought. In this view, language is a central ingredient of human thought as an adaptation to social modes of thinking that had evolved independently. This theme is elaborated in parts 2 and 3.

2

Language and Recursion

In 1871, Charles Darwin published *The Descent of Man in Relation to Sex*, in which he found the courage to declare that humans were descended from African apes. Just two years later, Friedrich Max Müller, who held the chair of philology at the University of Oxford, took exception:

> There is one difficulty which Mr. Darwin has not sufficiently appreciated. . . . There is between the whole animal kingdom on the one side, and man, even in his lowest state, on the other, a barrier which no animal has ever crossed, and that barrier is—*Language.* . . . If we removed the name of specific differences from our philosophical dictionaries, I should still hold that nothing deserves the name of man except what is able to speak . . . a speaking elephant or an elephantine speaker could never be called an elephant. Professor Schleicher, though an enthusiastic admirer of Darwin, observed once jokingly, but not without a deep meaning, "If a pig were to say to me, 'I am a pig' it would ipso facto cease to be a pig."[2]

Müller wrote in defiance of the ban on discussions of the evolution of language, imposed in 1866 by the Linguistic Society of Paris, and shortly afterwards by the Philological Society of London, but discussion of whether animal communication bore any resemblance to speech must have continued, even if surreptitiously. In 1919, Samuel Butler, the novelist, philosopher, and one-time New Zealand farmer, wrote:

> In his latest article . . . Prof Garner says that the chatter of monkeys is not meaningless, but that they are conveying ideas to one another. This seems to me hazardous. The monkeys might with equal justice conclude that in our magazine articles, or literary and artistic criticisms, we are not chattering idly but are conveying ideas to one another.[3]

The evidence suggests, though, that this is overly generous to monkeys and unkind to humans—although it may be accurate with respect to literary critics, even before postmodernism. So what is it that distinguishes human language from all other forms of animal communication?

Language as Recursion

As anticipated in chapter 1, a common answer is recursion, which provides for what Chomsky once referred to as the *generativity* of language.[4] Mark Hauser, Noam Chomsky and Tecumseh Fitch, in an influential article, have described recursion as the "minimum characteristic" distinguishing human language from animal communication.[5] The capacity to embed structures within structures in recursive fashion has endowed our species with a limitless capacity to create sentences to express an equally unbounded set of possible meanings. At least within the limits of one's memory and processing capacity, we can combine phrases to make sentences as long and complex as we like. The well-known children's story, *The House That Jack Built*, again provides a useful example:

> *This is the house that Jack built.*
> *This is the malt that lay in the house that Jack built.*
> *This is the rat that ate the malt that lay in the house that Jack built.*
> *This is the cat that killed the rat that ate the malt that lay in the house that Jack built.*

And so on . . . and on. Young children quickly understand that the sentence can be extended *ad infinitum*, or at least until you run out of breath. The recursive rules of grammar also allow phrases to be moved around instead of simply being tacked on to the beginning or end. The most demanding kind of recursion is what is called *center-embedded* recursion, in which phrases are embedded within phrases, instead of being tacked on. For example, if one wanted to highlight the *malt* in the story, one could embed phrases as follows:

> *The malt that the rat that the cat killed ate lay in the house that Jack built.*

This may require a moment or two to unpack. Too much concatenization like this can boggle the mind, probably because center-embedded recursion requires one to keep place markers where each phrase is interrupted by an embedded one. This places a strain on memory. In fact, sentences with more than one level of center-embedding (the above sentence has three) are extremely rare, and considered by some to be incomprehensible.[6]

The recursive nature of grammar can be expressed more formally in the so-called *rewrite rules* that specify how grammatical sentences are formed. As the examples from *The House That Jack Built* illustrate, sentences (S) can be constructed from phrases (P), which are then combined in recursive fashion. Three kinds of phrases are noun phrases (NP), verb phrases (VP), and prepositional phrases (PP). On a visit to a publishing house in Hove, England, I was once greeted by the publisher with the unlikely sentence *Ribena is trickling down the chandeliers*.[7] (It was.) Here, the sentence can be broken down into a NP (*Ribena*) and a VP (*is trickling down the chandeliers*). But the VP is itself composed of a verb (*is trickling*) plus a PP (*down the chandeliers*), which in turn is composed of a preposition (*down*) plus a NP (*the chandeliers*).

The recursion is most apparent if these relations are expressed as rewrite rules:[8]

1. S → NP VP
2. NP → article noun PP
3. VP → verb PP
4. PP → preposition NP

Here we see that the same phrasal elements can appear on either side of the rules that generate them. For example, a NP includes an optional PP, which in turn includes a NP. In principle, then, one could cycle repeatedly through rules 2 and 4. Had the publisher not been in something of a panic, he might have elaborated: *Ribena is trickling down the chandeliers onto the carpet beside my desk*. (By the way, there was a children's crèche upstairs.)

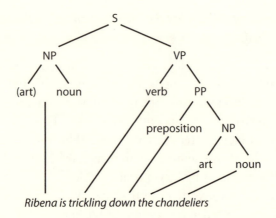
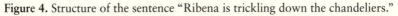

Ribena is trickling down the chandeliers

Figure 4. Structure of the sentence "Ribena is trickling down the chandeliers."

The structure of the sentence is also shown schematically in figure 4, which shows more clearly its hierarchical structure.

Once linguistic structures are established, we can then use language to refer to language, at another level of recursion. Consider the following sentence:

This sentence is a sentence.

As we may remember from school days, a sentence needs a verb, and this one has the verb *is* so the sentence is indeed a sentence, and is therefore true. But we can take that sentence and embed it in another sentence, as follows:

"This sentence is a sentence" is a sentence.

As it turns out, this is also true. And of course we could continue to embed each sentence so formed into the format *X is a sentence,* ad infinitum.

Sentences are the basic units of language that allow us to express propositions about the world, and propositions are in turn descriptions about states or actions that have so-called truth-value, which is to say that they are either true or false. Correspondingly, the sentences that express propositions are generally also either true or false, which gives humans the possibly unique privilege of being able to lie. There are a few tricky sentences, though, that cannot be declared either true or false, such as the enigmatic *This sentence is false*. It cannot be true, for if it were it would be false, and if it were false it would be true, if you catch my drift. Sentences like this have greatly exercised philosophers and logicians, from Eubulides of Miletus in the fourth century BC to Bertrand Russell and others in the twentieth century, and we are best to leave it to them to sort them out.

Chomsky's View of Language

The rewrite rules above are specific to a particular language, English. The rules are different for other languages, such as Chinese or Maori. Noam Chomsky has sought deeper rules that would apply to all languages. These rules are known as *universal grammar*. In the previous chapter I briefly referred to the most recent theory, known as the Minimalist Program,[9] according to which language, at its most fundamental level, can be reduced to a single operation, which Chomsky calls unbounded Merge. It is "unbounded" in the sense that it can be applied recursively, so that merged entities can themselves be merged to build up any desired level of complexity. The above examples illustrate how words can be merged into phrases, and phrases into sentences. Phrases can also be merged to form more complex phrases. To expand on the example used by Tecumseh Fitch, and mentioned in the previous chapter, articles such as *a, the, this, that*, etc, can be merged with nouns, such as *cat, dog, tree, lake*, and so on, to create noun phrases, such as *a dog, that dog, the lake, this tree*, and so on. These can be merged with prepositions, such as *by, near, beside*, and so on, to create more complex noun phrases, such as *near the tree, the dog beside the lake, a cat by that tree*, and so on.

The Merge operation, though, strictly holds for what Chomsky calls *I-language*, which is the internal language of thought, and need not apply directly to *E-language*, which is external language as actually spoken or signed. In mapping I-language to E-language, various supplementary principles are needed. For instance, in the example given in chapter 1, the merging of *Jane loves John* with

Jane flies airplanes to get *Jane, who flies airplanes, loves John* requires extra rules to introduce the word *who* and delete one copy of the word *Jane*. I-language will map onto different E-languages in different ways. Chomsky's notion of unbounded Merge, recursively applied, is therefore essentially an idealization, inferred from the study of external languages, but is not itself directly observable.

Because I-language is assumed to be the basis of all language, Chomsky argues that it must have no external reference, and therefore cannot have evolved through natural selection. Instead, it must have emerged in a single step, perhaps a mutation, probably within the past 100,000 years. He writes as follows:

> Within some small group from which we are descended, a rewiring of the brain took place in some individual, call him *Prometheus*, yielding the operation of unbounded Merge, applying to concepts with intricate (and little understood) properties. . . . Prometheus's language provides him with an infinite array of structured expressions.[10]

The idea that the basis for language emerged in a single step in a single individual is remarkable, and smacks of the miraculous. It is nevertheless quite widely accepted that language evolved recently and was unique to *Homo sapiens*, and was perhaps even the principal defining characteristic of our species.[11] The question of how language evolved is considered more fully in chapter 4.

Although I-language is not directly observable, Chomsky has nevertheless felt confident in deriving its principles from observation of a single E-language. He was once quoted as saying:

> I have not hesitated to propose a general principle of linguistic structure on the basis of a single language. The inference is legitimate, on the assumption that humans are not specifically adapted to learn one rather than another human language. . . . Assuming that the genetically determined language faculty is a common human possession, we may conclude that a principle of language is universal if we are led to postulate it as a "precondition" for the acquisition of a single language.[12]

I-language is in essence the basis of universal grammar—the set of principles thought to underlie all languages.[13] Universal grammar has come under increasing attack in recent years, in part because of the sheer variety of languages and the rapidity with which they change.[14] Michael Tomasello, for example, recently declared "Universal grammar is dead." From a Chomskyan perspective, though, criticisms of universal grammar often miss their target, since individual languages are at a step removed from universal grammar itself. Nevertheless, these criticisms certainly challenge the view that the universal principles of language can be derived from a single language. Moreover, the sheer variety of human languages may threaten the view that I-language can be said to exist in anything like the form proposed by Chomsky.

To illustrate, let's consider one example that seems about as remote from English as it is possible to get.

The Pirahã

As young missionaries, Daniel L. Everett and his wife went to Brazil in 1977 with the aim of converting a remote Brazilian tribe known as the Pirahã (pronounced roughly as *peed'a-han*) to Christianity. Their aim was to learn the language to the point that they could translate the Bible, and so expose the Pirahã to the teachings of Christ. The language was so impenetrable to foreigners that earlier missionaries had been unable to learn it, but the Everetts lived among the Pirahã for six years, and Daniel Everett did succeed in acquiring their language.[15]

In the course of his stay there, he began to have doubts about religion, and eventually became an atheist. Everett's religious doubt appears to have stemmed from his discovery that the Pirahã have little sense of time, and live essentially in the present. They have no fiction or creation myths, or any sense of history, which were no doubt formidable barriers to the understanding of Christianity, or indeed any religion. Everett's interests shifted to linguistics, and at the time of writing he is a professor at the State University of Illinois at Normal.

Although the language was at first impenetrable, Everett also discovered it to be highly simplified in many ways, at least in terms of grammar and vocabulary—although it is rich in morphology and prosody.[16] It has no words for colors, and no words for numbers except for words that might be roughly translated as *one*, *two*, and *many*. There are no tenses other than a simple distinction between *present* and *not-present*, reflecting the fact that the Pirahã seem to live in the present, with little understanding of past or future. Of special interest, though, is Everett's claim that the Pirahã language has no embedding of phrases, and indeed no recursion. The Pirahã have remained monolingual despite more than 200 years of trading with Portuguese-speaking Brazilians and speakers of other native languages. One might be tempted to think that the Pirahã suffer from some genetic defect, but this is firmly rejected by Everett, who knows them well.

No doubt a child born to Pirahã parents, but raised in Boston, would have no special difficulty learning Bostonian English. Everett also points out that the Pirahã use sentences without embedded phrases to say things that in other languages would be expressed in sentences with embedding. That is, they have thoughts involving recursion, but use nonrecursive language to express them. For example, the Pirahã have no verbs like *say*, *want*, or *think*, which in English are normally used with embedded clauses, as in *I said that John intends to leave*. In Pirahã this would be expressed as the equivalent of *My saying John intend-leaves*.[17]

In describing the grammatical paucity of Pirahã language, Everett is adamant that he is not casting any aspersions on their intelligence. "I am *not*," he writes, "making a claim about Pirahã conceptual abilities but about their expression of certain concepts linguistically, and this is a crucial difference."[18]

Linguistic Diversity

Everett's claims are understandably controversial.[19] Nevertheless it is perhaps unlikely that the language of the Pirahã is particularly unusual, and languages from other oral cultures may well exhibit similar features. For example, the Iatmul language of New Guinea is also said to have no recursion.[20] Nicholas Evans gives another example from Bininj Gun-Wok, the term given to a group of dialects spoken in Western Arnhem Land in Australia. In that language, the English sentence *They stood watching us fight* is rendered as the equivalent of *They stood / they were watching us / we were fighting.*[21]

Linguists are becoming increasingly aware of the enormous diversity and sheer number of languages in non-Western and preliterate societies, to the point that universal grammar may be as much threatened with extinction as are the languages that threaten it.[22] For example, the 10 million or so people who inhabit the island of New Guinea and its Melanesian surrounds speak some 1,150 languages, which amounts to just under 10,000 people per language. In Vanuatu, with a total population of 195,000 people, 105 different languages have been identified, with an average of less than 2,000 speakers per language. Australia boasts countless different indigenous languages, and the people of Arnhem Land are highly multilingual, often speaking as many as six languages by the time they are adults. According to one estimate, 17 countries hold 60 percent of all languages, but make up only 27 percent of the world's population and 9 percent of its land area.[23] Many of these languages are dying out, but they probably give a much better appreciation of the nature of language as it emerged in hunter-gatherer societies than as it has developed in Boston.

Nicholas Evans suggests that the diversity of language was not driven by geographic isolation, since they coexist in areas where there are no geographic barriers, and several different languages are often spoken within the same household. Language variation also seems to be driven by more than random drift. Evans suggests that language serves as a kind of passport, marking the right to belong to a particular local society. As groups split off, they may make deliberate moves to differentiate their languages. For example, in the Uisia dialect of the Buin language on Bougainville Island the gender agreements have all been reversed relative to those in the other dialects. All the masculine words have become feminine and all the feminine words masculine. This seems clearly a deliberate move to differentiate a particular subcommunity from the rest.[24]

Even the parts of speech, the building blocks of theories of grammar, may not be universal. Nicholas Evans points out that there are languages without prepositions, adjectives, articles, or adverbs, and that there is no consensus among linguists on whether all languages even distinguish between nouns and verbs. Even if they do, it is not clear which words belong in which category. Evans gives the examples of *paternal aunt*, which is expressed by a verb in the Australian aboriginal language Ilgar; *know*, which is an adjective in Australian Kayardild; and *love*, which is simply a suffix in the South American language Tiriyo.[25] In defense of universal grammar, the psychologists Steven Pinker and Paul Bloom have claimed that "no language uses noun affixes to express tense."[26] But the language of the Kayardild does; it marks the past tense by adding the suffix *-arra* to the verb as well as the suffix *-na* to the noun that is the object of the verb.[20] Such diversity imposes severe strain on any attempt to discover a coherent universal grammar.

Michael Tomasello concurs that "there are very few if any specific grammatical constructions or markers that are universally present in all languages."[27] He suggests that theories of language have been unduly influenced by the characteristics of written language, and are therefore confined to only a tiny fraction of the world's languages. Literacy emerged long after language itself, and is still far from universal. We should therefore not be contemptuous of the Pirahã, or indeed of other cultures relying on oral or signed communication and having no tradition of literacy. As mentioned above, even speakers of Western languages rarely use center-embedding in speech; an analysis of one corpus showed that 96 percent of *that* clauses, as illustrated in *The House That Jack Built*, were end-embedded and only 3 percent were center-embedded. The remaining 1 percent were initially embedded (as in *That Jack built the house is not disputed*).[28] Written language is more tolerant of multiple center-embedding, perhaps because the sentences remain in front of us while we try to process them. The ancient Greeks and Romans may have been partly to blame. Aristotle laid down the rules for the construction of sentences according to the doctrine of *periods*; a *periodic* sentence was defined as one with at least one center-embedding. The Latin scholars Cicero (106–43 BC) and

Livy (64 BC–AD 13) developed the periodic form, and their writing served as stylistic icons for centuries; their influence persists in the present-day languages of Europe.[29]

It should be noted that, in principle at least, Chomsky's approach does accommodate at least some degree of diversity in E-languages—the languages we actually speak or sign. The mapping from a universal I-language to E-language, he writes, "might turn out to be intricate, varied, and subject to accidental historical and cultural events: the Norman conquest, teen-age jargon, and so on."[30] Diversity can also arise in the way in which different cultures fit E-language to the constraints of speech. This is called *linearization*, since spoken language, at least, is constrained by the fact that words are spoken strictly in sequence, whereas I-language is unconstrained by linearization. The critical question, then, is not whether the Chomskyan view can accommodate linguistic diversity, but whether the degree of diversity recorded by authors such as Evans and Levinson could have arisen in so short a time. What seems more likely is that grammar itself evolved gradually rather than as a singular event within the past 100,000 years. Some views on how this might have happened are captured in the concept of *grammaticalization*.

Grammaticalization

If grammar does not depend on some inborn, universal set of principles, what does it depend on? Grammaticalization is the view that it emerged through a gradual process, driven more by practical concerns than by any biological predisposition.[31]

One of the processes involved in grammaticalization has to do with the changing role of words, leading to more efficient and economical expression.[32] For example, many of the words we use do not refer to actual content, but serve functions that are purely grammatical. These are called function words, and include articles, such as *a* and *the*, prepositions such as *at*, *on*, or *about*, and auxiliaries such as *will* in *They will come*. Function words nevertheless almost certainly have their origins in content words. A classic example is the word *have*, which progressed from a verb meaning to *seize* or *grasp* (Latin *capere*), to one expressing possession (as in *I have a pet porcupine*; Latin *habere*), to a marker of the perfect tense (*I have gone*) and a marker of obligation (*I have to go*). Similarly, the word *will* probably progressed from a verb (as in *Do what you will*) to a marker of the future tense (*They will laugh*).

Another example comes from the word *go*. It still carries the meaning of travel, or making a move from one location to another, but in sentences like *We're going to have lunch* it has been bleached of content and simply indicates the future. The phrase *going to* has been compressed into the form *gonna*, as in *We're gonna have lunch*, or even *I'm gonna go*. Some of my friends in the United States make an additional compression when they say *Let's go eat*, where we less hungry Kiwis say *Let's go and eat*. I understand that *Let's go go-go* is the battle song of the Chicago White Sox. This is language on the move—watch where *go*'s gonna go next.

There are other ways in which grammaticalization operates to make communication at once more streamlined and less error-prone. One has to do with the embedding and concatenization of phrases. For example, the statements *He pushed the door* and *The door opened* can be concatenated into *He pushed the door open*. Statements like *My uncle is generous with money* and *My uncle helped my sister out* can be concatenated by embedding the first in the second: *My uncle, who is generous with money, helped my sister out*. One can also alter the priority of the two statements by reversing the embedding: *My uncle, who helped my sister out, is generous with money*.

Efficiency can also be improved by breaking down concepts into component parts, which can then be recombined to form new concepts. An interesting example comes from a signed language. In Nicaragua deaf people were isolated from one another until the Sandinista government assumed power in 1979, and created the first schools for the deaf. Since that time, the children in these schools invented their own sign language, which has blended into the system now called Lenguaje de Signos Nicaragüense (NSL). In the course of time, NSL has changed from a system of holistic signs to a more combinatorial format. For example, one generation of children were told a story of a cat that swallowed a bowling ball, and then rolled down a steep street in a "waving, wobbling manner." The children were then asked to sign the motion. Some indicated the motion holistically, moving the hand downward in a waving motion. Others, however, segmented the motion into two signs, one representing downward motion and the other representing the waving motion, and this version increased after the first cohort of children had moved through the school.[33] These two signs can then be individually combined with other signs to create new meanings.

One need not appeal to universal grammar to explain how this kind of segmentation occurs. Computer simulations have shown that cultural transmission can change a language that begins with holistic units into one in which sequences of forms are combined to produce meanings that were earlier expressed holistically.[34]

In the Beginning Was the Word

According to the linguist Mark Aronoff, even words may have gained combinatorial structure over time. Words are generally considered to be made up of parts. At one level, they are composed of *phonemes*, the smallest units of speech or signed language that make a difference to meaning. In English, the words *cat* and *bat* differ only in the first phoneme, with corresponding differences in meaning, but those same phonemes /c/ and /b/ are used in combinations with each other and with other phonemes to create countless other words. The "phonemes" (earlier called *cheremes*) of American Sign Language are defined in terms of location, hand configuration, and movement, and similarly combine to form different signed words.[35] At the next level of structure, phonemes are themselves combined to form *morphemes*, which are the smallest units of meaning. Words like *jump* and *cat* are morphemes, but so are the endings of words that alter their grammatical status, such as the addition of *-ed* to signal the past tense, or the addition of *-s* to indicate the plural. Thus the words *jumped* and *cats* each consist of two morphemes. Other morphemes change the meaning of a word, such as the addition of the prefix *un-* to reverse the meaning, as in *happy* and *unhappy*.

Aronoff is among a group of linguists who have documented the emergence of ABSL, the signed language of the Al-Sayyid, a Bedouin community of some 3,500 individuals in the Negev Desert in Israel. Some 150 of the people living there have inherited a condition that has left them profoundly deaf. Although the deaf are in the minority, ABSL is widely used in the community, along with a spoken dialect of Arabic. It is a recent language, now in only its third generation of signers, and may be regarded as still in its infancy. Aronoff has remarked, though, that ABSL has no phonemes or morphemes. Each signed word is essentially a whole, not decomposable into parts.[36]

In this respect, ABSL seems to defy what has been called *duality of patterning*, the fact that language involves combinations of elements at two levels, the phonological and the grammatical. The linguist Charles F. Hockett listed duality of patterning as one of the "design features" of language,[37] so its absence in ABSL may be taken to mean that ABSL is not a true language. Yet Hockett himself recognized that the design features did not appear all at once, and that duality of patterning was probably a latecomer:

> There is excellent reason to believe that duality of patterning was the last property to be developed, because one can find little if any reason why a communication system should have this property unless it is highly complicated. If a vocal-auditory system comes to have a larger and larger number of distinct meaningful elements, those elements inevitably come to be more and more similar to one another in sound. There is a practical limit, for any species or any machine, to the number of distinct stimuli that can be discriminated, especially when the discriminations typically have to be made in noisy conditions.[38]

ABSL, then, may be regarded as a language in the early stages of development. Aronoff concludes that words are the primary elements, and do not acquire phonology or morphology until they are thrust upon them. As the title of one of his recent articles aptly puts it, "In the beginning was the word."[39]

Morphology may well arise with the compression of what were

separate words into compounds. For example, the addition of -ed to a verb to signal the past tense probably derives from the verb *to do*. Hence, roughly speaking, sentences like *He laughed* may have derived from something like *He laugh did*.[40] This follows a dictum stated by the functional linguist Talmy ("Tom") Givón that "Today's morphology is yesterday's syntax."[41]

The story that is beginning to emerge, then, is that language does not appear fully formed in different cultures as a product of universal grammar, but comes about gradually as a product of culture and accumulated experience, and a practical concern to make communication more efficient. That is, it grammaticalizes itself. It adjusts in somewhat the same way as my computer system for filing reprints has adjusted, with folders eventually gaining folders, and these in turn gaining further embedded folders. As my collection of reprints grows, my folders have embedded folders thrust upon them.

This does not mean that language has no genetic component—it is, after all, restricted to our own species, as I shall argue in the next chapter. It remains something of an open question just how much of language depends on innate components specific to language itself, and how much on more general aspects of the human mind. It may depend not so much on what Steven Pinker, echoing Chomsky's notion of universal grammar, called "the language instinct"[42] as on what has been termed an "instinct for inventiveness,"[43] coupled with a drive toward increased efficiency, that covers many other aspects of our lives, including art, music, and machines—not to mention filing systems.

What of Recursion, Then?

Languages, then, appear to have evolved gradually, adapting themselves to particular cultures, and undergoing progressive modifications. In the case of the Pirahã, for example, their lack of concern for time seems to have been critical. Everett writes:

> [The] apparently disjointed facts about the Pirahã language—gaps that are very surprising from just about any grammarian's perspective—

ultimately derive from a single cultural constraint in Pirahã, namely, *the restriction of communication to the immediate experience of the interlocutors.*[44]

Morten Christiansen and Nick Chater, in an influential article,[45] suggest that the sheer diversity of languages, and the rapidity with which they change, mean that language adapted to the brain, rather than the brain adapting to language. That adaptation, moreover, must have depended on mental functions that were not themselves primarily linguistic, and that were heavily influenced by environmental factors such as culture and geography.

The nature of those thought processes is explored in later chapters of this book. In particular, I will focus on mental time travel and theory of mind as nonlinguistic processes involving recursive processes that shaped how language evolved. Whereas Chomsky viewed thought through the lens of language, I (and others) suggest that language should be viewed through the lens of thought. This can have a liberating effect on our understanding of language and its evolution.

First, language can be understood to have evolved gradually, rather than as having emerged suddenly in some comparatively recent individual on the family tree, called Prometheus. We can suppose instead that language adapted to the cultural and practical demands faced by our forbears over several millennia, rather than just over the past 100,000 years. Prometheus, in short, can be unbound.

Second, any universal principles underlying language can be regarded as principles of human thought, and not specific to language. Even recursion appears not to be universal, and may be absent in many indigenous languages. There is therefore some doubt about whether it is truly the "minimum characteristic" that distinguishes human language from animal communication, as recently claimed.[46] The fact that many, perhaps most, human languages use recursion is of course good evidence that the human mind is capable of it, even if language can generate any number of different meanings without it. In oral societies such as the Pirahã, word combinations and repetition of phrases may be sufficient to generate all of the meanings required without recourse to the embedding of phrases or other recursive devices. And as Everett pointed out, there is no question that the Pirahã are capable of recursive *thought*.

It is perhaps ironic that a human capacity often said to depend on recursion can function without it, at least in the sense of phrases being embedded in phrases—languages like Pirahã may well be recursive in that the I-language underlying them may depend on a recursive operation, such as Chomsky's Merge. Moreover, language might still depend on the recursive nature of theory of mind, as I argue in chapter 9. It might be suggested that recursion is merely part of a toolkit for constructing language, and that not all languages use all of the tools.[47] A universal toolkit, though, is not quite the same as universal grammar. What the use of recursion in language does illustrate, though, is that recursion is a property of the human mind, employed when needed. As we shall see in later chapters, it is not specific to language.

That said, it remains highly likely that language itself is specific to humans. We have a compulsion to talk, or develop signing systems if speech is prevented, and children do seem to go through preprogrammed steps in learning language. The emergence of language in our species probably depended on the evolution of intentional systems of thought, and the adaptiveness of sharing our thoughts. The communicative imperative also drove anatomical changes, such as the redesign of the vocal tract to permit greater variety of sounds, and perhaps changes in breathing and intentional control.

These matters are discussed in later chapters. The more immediate matter of concern is whether animals other than humans can be said to possess anything resembling human language.

3

Do Animals Have Language?

The least thing upset him on the links. He missed short putts
because of the uproar of the butterflies in the adjoining meadows.
—from *The Clicking of Cuthbert*, by P. G. Wodehouse

Having grown up on a farm, I suspect that the above quotation is
an unwarranted slur on the gentle butterfly, but vocalization is ubiquitous among animals, including insects—and of course ourselves.
It has therefore seemed natural to suppose that human language
must have evolved from animal calls. Only in fiction, though, do
animals actually speak and hold meaningful conversations. Most
examples, from *Winnie the Pooh* to the Beatrix Potter books, are
written for children, but my favorite example comes from a short
story entitled *Tobermory*, by Saki, who made a brief appearance
in the previous chapter. Tobermory is a cat who has been taught
to speak, to the consternation of guests at a house party. Here is
what Tobermory says to a woman who is foolish enough to ask his
opinion of her intelligence:

> "You put me in an embarrassing position," said Tobermory, whose
> tone and attitude certainly did not suggest a shred of embarrassment.
> "When your inclusion in the house party was suggested Sir Wilfred
> protested that you were the most brainless woman of his acquain-
> tance, and that there was a wide distinction between hospitality and
> the care of the feeble-minded. Lady Blemley replied that your lack of
> brain-power was the precise quality which had earned you your invi-
> tation, as you were the only person she could think of who might be
> idiotic enough to buy their old car."[1]

To everyone's relief, Tobermory was killed shortly afterwards by a
tomcat from the Rectory.

In most respects, though, animal vocalizations have little in common with human speech. For the most part, they are genetically fixed, and under emotional rather than voluntary control. They are organized in the midbrain, with little or no input from the cortex, the seat of higher mental functions. For example, electrical stimulation of a region of the midbrain called the periaqueductal grey induces hissing, growling, screaming, howling, and meowing in the cat, shrieking, yelling, yapping, cackling, and trilling in the squirrel monkey, echolocation sounds in the bat, and laughing in humans. These vocalizations seem to be organized downstream from the periaqueductal grey, in the ancient depths of the brain.[2] Thus, destruction of the periaqueductal grey causes rats, cats, squirrel monkeys, and humans to become mute.

Human laughter, then, belongs in the category of innate vocalizations unrelated to speech, and is correspondingly difficult to produce voluntarily or to suppress. Robert Provine, in his book *Laughter: A Scientific Investigation*, records that a girls' boarding school in Tanzania had to be closed because of an uncontrollable epidemic of hysterical laughter.[3] And conversely, it takes a trained actor to laugh convincingly in the absence of the emotional state, although my father, being Irish, was able to laugh uproariously at incomprehensible jokes told by his mates. This only led to embarrassment when others phoned later to ask the point of the joke, and he had to confess that he didn't know.

Even the calls of chimpanzees, our closest living relatives, appear to be largely involuntary. Jane Goodall, who lived among the chimpanzees in the wild at Gombe National Park in Tanzania, once recorded the instance of a young chimp that had discovered a cache of bananas, and evidently wanted to keep them for himself. But when chimpanzees discover food, they characteristically emit a pant hoot call, drawing the attention of other members in the troop. The young chimpanzee was unable to suppress the emotionally driven call, but did the best he could to muffle it by placing a hand over his mouth. Conversely, chimpanzees find it difficult to produce a call voluntarily. Goodall remarks, "The production of sound in the absence of the appropriate emotional state seems to be almost an impossible task for a chimpanzee."[4]

This may nevertheless be something of an exaggeration. Different pant hoot calls have been recorded in different regions of Africa, suggesting a degree of cultural influence. But even here the variation may not be so much in the actual sounds as in their timing.[5] Pant hoots are often accompanied by drumming, in which the animals repeatedly hit their hands or feet against a surface—including their own chests. Groups have a characteristic beat, and the resulting differences may serve as identification tags, allowing the animals to maintain contact. Chest beating, along with threatening movements and vocalizations, is also well documented in mountain gorillas.[6] These repetitive vocalizations and drumming may explain the origins of rock concerts, rather than of speech.

Chimpanzees also modify their calls in other ways. When under attack, chimps scream, and one study has shown that their screams have different acoustic structure if there is a listener in the audience of higher rank than the aggressor.[7] In another study, captive chimps were more likely to produce two attention-getting sounds, the "raspberry" and the "extended grunt," when a human appeared with a favorite food than when either the human or the food appeared alone.[8] Such observations suggest a limited degree of voluntary control, although these variations may also be under subtle emotional control. Perhaps it's a bit like the limited way we can modify emotional sounds such as laughing or crying, depending on the audience, or on the status of the person we're responding to.

In some respects, chimpanzees and other great apes, despite being our closest living relatives among the primates, may not provide a good primate model for the voluntary control of vocalization.[9] Charles T. Snowdon somewhat bemusedly refers to "the silence of the apes";[10] the great apes, he says, simply don't vocalize much at all, at least in comparison with other primates, including ourselves. The primates of the Amazon, in contrast, join in the frenzy of sound created by frogs, birds, and other creatures in the rain forest, where vocalization presumably acts as a signal to distinguish one species from another. Snowdon claims to be able to locate a group of pygmy marmosets, the world's smallest monkeys, simply by listening for their distinctive and persistent vocalizations. The mountain gorillas of Rwanda, on the other hand, are especially conspicuous for their silence, at least according to Snowdon.[11] Maybe they are just strong, silent types, more into chest-beating displays than flights of vocal eloquence. And maybe, too, it's sometimes the strong, silent, Tarzan type, rather than the rock star, who gets the girl.[12]

There may also be better voluntary control of vocalization in primates more distantly related to humans than are the great apes. Macaques can learn to increase the rate of vocalization to gain food reward or avoid shock, but this is abolished by bilateral destruction of a cortical area known as the anterior cingulate.[13] In squirrel monkeys, destruction of this area abolishes spontaneously uttered long-distance calls, but leaves intact the animals' ability to respond with contact calls to contact calls made by others.[14] In these cases, the calls themselves are innate, but their production appears to be under some intentional control, mediated by the anterior cingulate. Japanese monkeys have been taught to emit "coo" calls to make requests, also suggesting a degree of voluntary control. More provocatively, it is claimed that they can modify their coos in order to receive either food or tools, suggesting that they may have been able to bring their vocalizations partly under the control of the motor cortex.[15] It has also been suggested, though, that the coo sound was entirely involuntary, the result of an unconscious link, well documented in humans as well as primates, between limb movements and vocalization.[16]

Human speech is not only intentional, it also requires the learning of new sound patterns. Although most animals and many birds vocalize, very few are capable of vocal learning. Just why this is so is something of a mystery. Erich D. Jarvis suggests that there has been selection against vocal learning because it introduces variation, and makes the calls more noticeable to predators—just as we humans tend to notice the sound of a new voice. Jarvis suggests that animals that do learn new patterns are the ones without major predators, with humans at the top of the list. Apart from humans, killer whales are the top predators of the ocean, and are vocal learners. Adult elephants are also vocal learners, and do not have natural predators, although lions, hyenas, and crocodiles sometimes prey on young elephants. Another vocal learner, the hummingbird, is said to be fearless, because it can easily escape predation through rapid flight, and songbirds such as parrots and ravens can evade predation through effective mobbing behavior.[17] One may wonder, though, why the dominant cats of the African savannah, such as the lion (king of the jungle), are not vocal learners. Perhaps they are, and no one has dared get close enough to find out. Or maybe they learned to keep quiet, and so escape notice, when humans developed the facility to kill them (and everything else).

Another suggestion is that innately programmed vocalizations can't be faked, and can therefore be trusted.[18] If an animal cries "wolf" it is important that a wolf be actually present, otherwise her mates may not believe her next time and fall victim to that voracious animal. Human talk, in contrast, is notoriously untrustworthy, as the poet Robert Graves warns in his poem *Beware Madam!*

> Beware, madam, of the witty devil
> The arch intriguer who walks disguised
> In a poet's cloak, his gay tongue oozing evil.

The idea that animal communication cannot be faked has been challenged, though,[19] and we might at least make an exception of birds, some of whom sing to deceive by imitating other birds, or even humans. The lyre bird in Australia is said to be able to imitate the sound of a beer can being opened, although with wine having largely displaced beer as the national drink in that country we might expect these birds soon to imitate the popping of corks.[20]

If animals aren't conversing with one another, one may well ask what all the noise is about. Usually, it has to do with essentially instinctive or emotional situations such as mating, aggression, territorial claim, or warning of predators. One of the most studied examples is the vervet monkey, which has separate cries to warn of the presence of a snake, an eagle, a leopard, a smaller cat, and a baboon, and when monkeys hear these calls they act appropriately—running up into the trees, for example, in response to the leopard call.[21] Even those birds that learn new patterns of sound appear not to be conversing, or conveying new information. Erich Jarvis suggests that vocal learning evolved in the first instance not to transmit meaning, but rather to defend territory[22] and attract mates. In humans, as in songbirds, singing is an activity that attracts potential mates and allows individuals to establish themselves as sex icons. I have a colleague who once traveled a considerable distance to join a queue to be kissed by Elvis.[23] This raises the intriguing possibility that speech evolved from singing, an idea explored by Steven Mithen in his 2005 book *The Singing Neanderthals*.[24]

Vocal learning is often employed in order to imitate, as in the case of the lyre bird. Imitation presumably enables the imitator to encroach on the territory of another species, and take advantage of their resources. Some animals have even proven to be adept at imitating human speech. In his 1997 book *The Symbolic Species*, Terrence Deacon records his astonishment when, as he was passing in front of the Boston Aquarium, a voice called out "Hey! Hey! Get outa there!"[25] It came from Hoover, the now legendary talking seal, who died, sadly, in 1986. An elephant named Kosik at a South Korean Zoo has recently been recorded uttering several Korean words and phrases. The vocal range of elephants is normally too low for humans to hear, but Kosik has found a way to curl his trunk into his mouth and blow into it, and so create frequencies high enough to produce sounds recognizable as speech.[26] But simple imitation is not the same as speech produced through the use of grammatical rules. Closer to the mark was Alex, an African gray parrot, who also died recently. Alex's voice has been described as like a recording from an old-style Victrola.[27] He did not merely mimic, but was able to answer simple questions about the colors or shapes of objects, or about the number of objects up to about six. He was said to have had the speech capabilities of a two-year-old human child.[28] This is progress indeed, but just as the two-year-old has not yet developed true grammatical speech, neither did Alex.

For the birds, I should also mention another African gray parrot called N'kisi, and featured in a BBC News report of 26 January, 2004. N'kisi belongs to Aimee Morgana (or her to him), and is said to have a vocabulary of 950 words and to generate novel utterances. He appears to have a mischievous sense of humor. On one occasion he met Dr. Jane Goodall, well known for having lived among, studied, and befriended chimpanzees in the wild. Having previously seen her in a picture with chimpanzees, N'kisi asked, "Got a chimp?" He apparently has some understanding of grammar, and it is claimed that he once said "flied" for "flew"—the kind of error often made by young children who have learned morphological rules but not yet learned the exceptions. More dramatically still, it is claimed that N'kisi has the power of telepathy, and Aimee Morgana has teamed with Dr. Rupert Sheldrake, whose most recent book is entitled *Dogs That Know When Their Owners Are Coming Home*, to demonstrate this power. N'kisi, they say, can telepathically surf the leading edge of Aimee's consciousness.

I relate this, not to argue that parrots possess language, or to convince you that telepathy exists, even if restricted to dogs and parrots. Rather, claims of this sort appear all the time, but nearly always prove to be without solid foundation. As far as I know, N'kisi's exploits have yet to be subjected to rigorous scientific testing, or published in a reputable scientific journal.[29] Even so, a closed mind on the subject—as on any subject—is not recommended, and may deny your parrot, or your dog, the chance to surf.

And we do need to be especially careful in drawing inferences about the mental capacities of nonhuman animals, especially in light of the famous 1904 case of Clever Hans, a horse. According to his trainer, Clever Hans could answer complex questions by tapping out letters of the alphabet with a front hoof, with each letter represented by a different number of taps. When asked "What is two-fifths plus one-half?" he stamped his hoof nine times, paused, and stamped another 10 times, apparently indicating that the answer was nine-tenths. Even the leading psychologist of the day, Professor Stumpf of the University of Berlin, was convinced of the horse's genius, until one of his students, Oskar Pfungst, showed that Clever Hans was actually responding to subtle signals given by his trainer. The trainer himself apparently did not realize that he, and not Clever Hans, was generating the answers.

What, then, of our closest relatives, the chimpanzee and bonobo?[30] Even at the acoustic level, without any consideration of meaning, the vocal exchanges between chimpanzees differ markedly from the exchanges that occur in human conversation. When people converse, they generally choose words, and therefore sounds, that are different from those they have just heard—the answer to a question, for example, is not the same as the question itself, unless perhaps it comes from a postmodernist, or a psychoanalyst. The sounds that chimpanzees emit during vocal exchanges are similar to what they have just heard.[31] Their echoed exchanges probably have to do simply with maintaining contact, rather than with telling what Bobo said to Mimi last night over dinner.

Attempts to teach our great ape cousins to speak have been notoriously unsuccessful. The best-known attempt was that of Cathy and Keith Hayes, a husband-and-wife team, who raised a baby chimp called Viki in their own home, along with their own children, hoping that speech would come as naturally to Viki as to the other kids. But it did not. At best, Viki was able to produce crude, approximations to three or four words: *mama, papa, cup,* and possibly *up.*[32] These painful efforts were whispered rather than vocalized, suggesting that at least part of the problem lay in the vocal component itself. But even the whispered attempts were far removed from the effortless, articulate, but often infuriating chatter of a bright three-year-old human. Chimps, then, seem to be distinctly at a loss for words. Parrots can do a much better job of articulating something like human speech than any primate can.

Another classic case is reported by the pioneering Russian psychologist Nadesha Ladygina-Kohts, who undertook detailed studies of a baby chimpanzee called Joni, and compared his development with that of her son Roodi.[33] Joni learned nothing approximating speech, but nevertheless loved to make sounds, including laughter. He also snored in a manner indistinguishable from that of a child. His vocalizations, though diverse, were attributable largely to his emotional state. Oddly, he could imitate the barking of a dog,[34] but not the sounds of human speech. Ladygina-Kohts grew almost as attached to Joni as to her own son, and her studies of Joni's behavior were meticulous and insightful, but she nevertheless concludes her book with a paean to human superiority, based largely on our unique language ability. After noting the "equalities" between child and chimp, she writes:

And, as soon as we take the child's verbal expression into consideration we immediately have to change these "equals" signs we were about to put between the intellectual capacities of a 4-year-old child and his chimpanzee counterpart. We change the equals sign into the sign >, which still does not look expressive enough; you want to say, or rather yell, not only more, but better, qualitatively higher, and incomparably, inexpressibly more perfect![35]

Do Animals Understand Us?

Although nonhuman animals have little or no ability to produce anything resembling human speech, they may have surprising ability to understand it. One of the more remarkable instances comes, not from an ape, but from a border collie.[36] His name is Rico, and he is able to respond accurately to spoken requests to fetch different objects from another room, and then to either place the designated object in a box or bring it to a particular person. In experimental trials, he was given 10 objects randomly selected from 200 objects that he knows, and chose correctly in 37 out of 40 trials. Rico collects the designated object from a room in which there is no person who might cue him about the correct selection, which rules out any "Clever Hans" effect. If he is given an unfamiliar name of an object to fetch, he will choose the one object among the otherwise familiar selection that is novel. Four weeks later, he demonstrates that he still knows the name of this object, indicating what has been termed "learning by exclusion." This ability to apply a label on a single trial is known as "fast mapping," and has hitherto been thought to be restricted to humans.[37] Rico's exploits may not come as a surprise to people convinced that their pet dogs or cats can understand them.

Rico's performance is somewhat comparable to that of Kanzi, a bonobo raised by Sue Savage-Rumbaugh. Kanzi is unable to speak, but as we shall see he has acquired an impressive facility to communicate by using manual gestures. What is interesting here is that his ability to understand spoken language far exceeds his ability to produce it. He can respond correctly to quite long sentences. For example, when asked, "Would you put some grapes in the swimming pool?" he immediately got out of the water, fetched some grapes, and tossed them into the pool. When visiting his friend Austin, a chimpanzee, he was told, "You can have some cereal if you give Austin your monster mask to play with." He responded by finding his mask and giving it to Austin, and then pointing to Austin's cereal. His ability to respond to such commands is not perfect, though. In one controlled study, he was given a list of 660 unusual spoken commands, some of them eight words long, and responded correctly on 72 percent of them. Kanzi was nine at the time, and did a little better than a two-and-a-half-year-old girl called Alia, who managed to get 66 percent correct.[38]

These examples suggest that comprehension of speech far outstrips production, also a common observation in children acquiring language.[39] They also suggest a surprising ability to break sentences down into words, hitherto considered a uniquely human capacity. Although this is something that seems natural to most of us, there is virtually nothing in the acoustic signal that tells us where one word ends and another begins,[40] and it is really only experience with the language that enables us to break a sentence down in this way.[41] Doyoufollowme? We become aware of this only when listening to a language that is completely foreign, when all the words seem to run together in a meaningless babble. When we teach children to speak, we help them to separate the words with an exaggerated form of speech known as "motherese."[42] The surprise, then, is not just that Rico and Kanzi were able to respond correctly to words, it is that they were able to pick them out at all. One must also suppose that they are not unique among their species—presumably, other apes and mammals are capable of the same thing, given the right conditions of learning. Keep talking to your cat.

It is unlikely, though, that the understanding exhibited by Rico and Kanzi—one hopes they might one day meet—meets the definition of true language comprehension. The identification of key words may be sufficient to provide the correct response most of the time. All Rico needs to know is the name of the object and the name of the recipient (box or person), and the rest follows.

Although the sentence "You can have some cereal if you give Austin your monster mask to play with" involves recursion (specifically end-recursion), Kanzi probably did not need to parse the sentence in order to understand what he needed to do. He just needed to pick out the words *cereal, Austin,* and *mask* to have a pretty good idea of what was required.[43]

The exploits of Rico and Kanzi nevertheless clearly depend on sophisticated analysis of acoustic signals, to a level well beyond the sounds they are actually able to produce. Although their calls seem to be largely fixed and under emotional control, animals hear a lot of different sounds in the wild, and need to be able to discriminate among them and at times take appropriate action. These sounds include the calls of other animals, including those of predatory humans. In the jungle, you need to stay tuned.

What about Signs?

Most nonhuman animals are highly vocal, but seem incapable of anything resembling human speech. The fact that some animals, at least, can understand speech suggests that any linguistic deficit may have to do with the production of intentional, learned vocal output, and not necessarily with language itself. Language, though, need not be vocal. The signed languages of the deaf are now clearly recognized as true languages, with an expressive power equal to that of speech—and in the next chapter I argue that human language evolved from manual gestures, and not from vocalizations. In literate societies reading and writing also comprise a form of language that can be accomplished without sound.[44]

A hint that it might be possible to teach visual language to apes came from an observation made by the English diarist Samuel Pepys. In 1661, he saw a strange creature, probably a chimpanzee or gorilla,[45] which had been brought from Guinea, and wrote that it was "so much like a man in most things . . . I cannot believe it is a monster got of a man and a she-baboon. I do believe it readily understands much English; and I am of a mind it might be taught to speak and make signs." He was wrong about speaking; as we have seen, there is no evidence whatsoever that any nonhuman primate can produce anything resembling human speech.

Pepys may have struck a seam of truth, though, in his suggestion that an ape might be taught to make signs. Primates, including apes, have hands and a manual control system that is well adapted for grasping and manipulating things, and is much better adapted to learning and intentional control than is their vocal system. This realization led Allen and Beatrix Gardner to develop a system of signs, based loosely on American Sign Language, in an attempt to communicate with a young chimpanzee called Washoe. This was indeed much more successful than earlier attempts to teach chimpanzees to speak. Washoe learned well over 100 gestures, in marked contrast to the mere three or four words that Viki had attempted to mouth. Washoe's first signed "word" was *more*, made by bringing the hands together in front of the body. She used this sign in combination with other signs to request more treats, such as bananas or tickles.[46] Later, Francine Patterson taught a gorilla named Koko at least 375 signs, and claimed that Koko could use these signs in inventive, human-like ways, to lie, swear, and pour scorn. Patterson even claimed to have measured Koko's IQ at somewhere between 84 and 95.[47]

The most impressive results so far, though, are those obtained by Sue Savage-Rumbaugh with the bonobo Kanzi, using an approach based more on reading than on signed language. Kanzi communicates by pointing to keys on a keyboard containing 256 symbols denoting objects and actions, and has supplemented these with manual gestures of his own devising. The symbols, known as *lexigrams*, are deliberately selected not to bear any physical resemblance to the objects or actions they represent. There also seems little doubt that Kanzi can produce novel sequences by pointing to combinations of lexigrams on the keyboard, although these combinations are meager compared to human language. His "utterances" are two- or three-word combinations, such as *hide peanut*, *chase you*, *hot water there*, or *child-side food surprise*.[48]

There has been considerable debate, though, about whether Kanzi has acquired true language, and the general consensus is that he has not, and that these simple combinations do not constitute

Figure 5. Panbanisha, Kanzi's adopted sister, communicating with lexigrams. Photo by Malcolm Linton.

true recursive grammar. Steven Pinker remarked that great apes, for all their linguistic exploits, "just don't 'get it.' "[49] Of course, this may reflect a desperate longing to cling to the notion of human superiority, but it is nevertheless clear that Kanzi has a long way to go, if he is to get there at all. At a recent conference on the evolution of language, held in Paris, it was suggested that Kanzi himself might appear at the following conference, two years later, and deliver a talk. Careful inspection of the presenters at that conference persuaded me that Kanzi was not one of them.[50]

Although Kanzi is perhaps the star among nonhuman animals in his linguistic prowess, others are not so far behind. These include other great apes (gorillas and orangutans),[51] dolphins,[52] sea lions, and Alex, the African gray parrot.[53] The linguist Derek Bickerton coined the term "protolanguage" to refer to the ability to form or understand simple combinations of symbols, but without grammar.[54] It has been claimed that this is the level of language attained, not only by the species mentioned above, but also by two-year-old children, speakers of pidgin languages, those with certain brain injuries preventing fluent speech, and drunken teenagers.[55]

In the child, grammar develops between the ages of about two and four. This stage is part of childhood, which may be unique to humans, and critical to the development of other aspects of thought peculiar to humans—a point elaborated in chapter 11. Bickerton has proposed that protolanguage is the platform upon which fully grammatical language was built in the course of evolution, a view reiterated by Ray Jackendoff in a recent landmark book.[56]

Another view of protolanguage, at least as manifest in nonhuman animals, is that it is simply a form of problem solving. Indeed the linguistic exploits of Kanzi and other apes seem comparable in many ways to the problem-solving activities of the chimpanzees described in the famous experiments of the German psychologist Wolfgang Köhler. When the chimps were presented with problems to solve, Köhler noted that the solution often seemed to come suddenly, as though through a flash of insight, and the solution often involved combining two objects, or two ideas, in a novel way. For example, the most "intelligent" of the chimps, Sultan, once figured out how to rake in food that was just out of reach by joining two pieces of bamboo together to make a rake long enough to reach the food. On another occasion, he used a small bamboo stick to rake in a longer one that was out of reach, and then used the longer stick to rake in food.[57] These acts seem little different in principle from the combining of communicative gestures, and many of the "requests" generated by the so-called linguistic apes are also produced with the aim of receiving food.[58]

More recently, Michael Tomasello has demonstrated similar problem-solving ability in chimpanzees, and made the interesting observation that they do not seem to learn by imitating others. When there were two ways to rake in food, chimps preferred to work out their own way of doing it, and were not influenced by having seen it done the other way. Children, in contrast, are more likely to copy what they have seen.[59] This apparent lack of imitative ability in the chimp may partially explain why they have not pressed on to Churchillian feats of oratory, or the building of jumbo jets. Daniel Povinelli has also explored the abilities of chimpanzees to solve mechanical problems, and seems more impressed by their obtuse lack of understanding of the physical world than by their occasional successes.[60] In the see-saw world of primatology, however, there is now evidence that chimpanzees may not only learn different ways of solving mechanical problems, but may transmit them to others in their group, suggesting a rudimentary basis for the establishment of culture.[61] It may be the chimpanzee's uncomfortable resemblance to ourselves that makes their mental abilities so much a matter of contention.

A curious feature of protolanguage as revealed by Kanzi and the other so-called linguistic apes is that there is little evidence for comparable communication among apes in the wild, or in naturalistic settings. Nevertheless they do gesture. Joanne Tanner and Richard Byrne counted some 30 different gestures made by lowland gorillas in the San Francisco Zoo, where the animals, cousins to those strong silent mountain gorillas we encountered earlier, are enclosed in a large, naturalistic setting. These gestures are closer to pantomime than language, though, and are easily understood by both human and gorilla observers.[62] Similarly, Michael Tomasello and his colleagues identified 30 gestures made by free-ranging chimpanzees at the Leipzig Zoo.[63] These gestures by no means exhausted the total repertoire of the animals, but were chosen because they could be readily observed and tabulated by the experimenters. These ape gestures, although unitary, are nevertheless somewhat language-like in that they are typically directed toward another individual, whereas animal calls tend to be directed to the community at large.

The manual gestures of chimpanzees and bonobos also differ from their facial movements and vocalizations in that they are less tied to typical contexts, such as play, grooming, or sex. They also show more variability between the two species and between subgroups within each species.[64] These features imply that manual gestures are used more freely and flexibly than are vocalizations, again suggesting that they are deployed intentionally, whereas vocalizations are largely under emotional control—on automatic pilot, as it were.

Michael Tomasello and his colleagues have also studied communicative bodily gestures made by great apes in the wild, and shown that these gestures are both subject to social learning and sensitive to the attentional state of the receiver.[65] These are both prerequisites for language—although they are not sufficient, as we shall see in chapter 9. In these studies gestures are defined as communicative movements of the head, limbs, or body, but exclude vocalizations. Some gestures, though, seem designed to produce sound—these include clapping and chest-beating. The gorilla appears to have a greater repertoire than the chimpanzee or bonobo, perhaps because the gorilla is the most terrestrial of the three, with arms less occupied with climbing and clinging.

But although ape gestures are clearly communicative and intentional, and subject to learning, they do not have the combinatorial generativity of human language. There are no sentences. Just what is missing from these gestures to enable them to develop into language is further explored in chapter 9.

Recursion in Nonhuman Species

Over 40 years ago, the distinguished linguist Noam Chomsky remarked that human language was "based on an entirely different principle" than all other forms of animal communication.[66] That conclusion seems much less true than it did then, and Chomsky himself appears to have softened his view. In an influential article he coauthored with Mark Hauser and Tecumseh Fitch, two definitions of the language faculty are proposed. *The faculty of language in the broad sense* (FLB) includes a great many features shared between humans and other species, including input-output systems, and what they call the "conceptual-intentional system," implying purposeful intention to communicate with others.[67] Contained within this system is the *faculty of language in the narrow sense* (FLN), which is in effect the I-language discussed in the previous chapter.[68] As we saw there, Chomsky argued that I-language emerged in a single step in late human evolution, and was thus clearly denied to nonhuman apes.

Hauser, Chomsky, and Fitch note that "all approaches [to language] agree that a core property of FLN is recursion."[69] In Chomsky's most recent Miminalist Program, it is the Merge operation that is applied recursively, as we saw in the previous chapter. Since Hauser and coauthors identified recursion as the critical component, there has been at least one claim that recursive parsing can be accomplished, not by apes, but by starlings.[70] It is worth taking a closer look, because any such claim poses a real challenge to our supposed uniqueness.

The claim was based on the parsing of center-embedded sequences, in which pairs of elements are progressively embedded in other pairs. The elements comprised rattles and warbles, the natural sounds that starlings make, and each pair was made up of a particular rattle and a particular warble. There were eight exemplars of each, so that the actual pairs were randomly chosen. An example of a center-embedded series with three levels of embedding would be $R_7R_1R_5R_3W_6W_7W_2W_5$—so that R_3W_6 is embedded in R_5W_7, and then this sequence is embedded in R_1W_2, and so on. Center-embedded sequences were compared with sequences made up of repeated pairs, as in $R_5W_1R_3W_2R_8W_5R_8W_4$. These structures are also shown in figure 6. After extensive training, the starlings were eventually able to discriminate both the center-embedded sequences and the repeated pairs from sequences not obeying these conditions.[71] Since the actual examples of rattles and warbles were varied randomly from trial to trial, the birds could not have been learning specific sequences, but must have somehow grasped the different structures.

Does that mean that the starling understood that the embedded sequence was made up of the recursive embedding of pairs within pairs? Alas no. The starlings might simply have been able to discern that the so-called embedded sequences comprised a sequence of rattles followed by an equal number of warbles. This does require the ability to estimate the number of each, at least up to four, and then judge whether the two numbers were the same. This structure is also shown in figure 6. To adopt this alternative strategy, they need some ability to count, or at least estimate number—birds are pretty good at that[72]—and some ability to judge whether two quantities are equal. They don't actually have to be terribly good at this, since their performance, although better than chance, was not perfect.

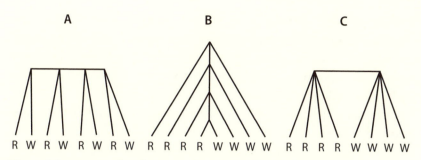

Figure 6. Sequences of rattles (R) and warbles (W) played to starlings.
A = repeated pairs; B = embedded pairs, C = alternative parsing that may have
allowed the starlings to distinguish B without appealing to recursive embedding.

It seems likely, then, that the starlings cleverly discovered a simpler solution to the problem set them, one that did not involve the understanding of recursion.[73] We might nevertheless be advised to pay more heed to the songs of starlings, perhaps following the lead of Wolfgang Amadeus Mozart. The final movement of his Piano Concerto in G Major is said to have been based on a song by his pet starling. More seriously, the foregoing example illustrates that the concept of recursion can be elusive, and one can be misled into thinking that a sequence that was constructed from recursive embedding is necessarily decomposed in the same manner.

If starlings were genuinely able to parse three levels of center-embedding, this would be deeply embarrassing for humans, who are pretty much incapable of understanding recursion at this level of complexity. For example, we can understand a single level of center-embedding, as in *John, whom Emily loves, adores Jane,* and at a pinch we can add another level, as in *John, whom Emily, whom Tom loves, loves, adores Jane.* But try adding another level: *John, whom Emily, whom Tom, whom Caroline loves, loves, loves, adores Jane.* That's not just an eternal triangle, it's an eternal tetrahedron.

For the present, then, we may safely conclude that there is no evidence for recursive parsing of sequences in nonhuman species. In language, at least, recursion may indeed be uniquely human, as Chomsky and others have maintained. Nevertheless we saw in the previous chapter that some human languages may not involve recursion, although that too may depend on how recursion is defined. In later chapters, though, I argue that human language would not have been possible were it not for the recursive nature of human thought.

I consider next the general question of how human language evolved.

4

How Language Evolved from Hand to Mouth

A tongue is a tongue
And a lung is a lung
In a tale you can shout or sing
Without the gesture? Nothing!
   —from "Gesticulate" in the 1953 musical *Kismet*

The 1866 ban by the Linguistic Society of Paris on all discussions of the evolution of language seems to have had a prolonged effect. The main difficulty, it seems, was (and to some extent still is) the widespread belief that language is uniquely human, so that there is no evidence to be gained from the study of nonhuman animals. This meant that language must have evolved some time since the split of the hominins from the great apes. In the nineteenth century, at least, there was little to be gleaned from fossil evidence, and any theory on how language evolved was largely a matter of speculation—and no doubt argument. Of course evolution was itself contentious, and was vigorously attacked by the church. In the case of language, the conflict between science and religion would have been exacerbated by the long-standing view that language was a gift from God.

Noam Chomsky's view of language, summarized in chapter 2, also leads to a somewhat miraculous view of how language evolved. The basis for all language, he proposes, is I-language, or the language of thought. Since I-language has no external referents, and must precede the evolution of E-languages—the languages we actually speak or sign—it cannot have arisen through natural selection. It must therefore have emerged as a singular event, in a single individual. In chapter 2, we were introduced to this individual as Prometheus. Chomsky also remarks, "Roughly 100,000+ years ago . . . there were no languages,"[1] which suggests that Prometheus lived after the actual emergence of our species, presumably in Africa.

The view that language emerged de novo in *Homo sapiens* has been elaborated by others, and is sometimes called the "big bang" theory. Derek Bickerton once wrote that "true language, via the emergence of syntax, was a catastrophic event, occurring within the first few generations of *Homo sapiens sapiens*."[2] The idea that language emerged very recently is based in part on the view that so-called modern human behavior, as inferred from the archaeological record and presumed to include language, arose within the past 100,000 years, and perhaps as recently as 50,000 years ago. Richard Klein, for example, writes that it becomes "at least plausible to tie the basic behavioral shift at 50 ka to a fortuitous mutation that created the fully modern brain."[3] Timothy Crow has proposed that a genetic mutation gave rise to the speciation of *Homo sapiens*, along with such uniquely human attributes as language, cerebral asymmetry, theory of mind, and a vulnerability to psychosis.[4] That was some bang.

From a Darwinian perspective this view is deeply implausible. Steven Pinker and Paul Bloom were perhaps the first to question the idea that language emerged in a singular event. Language, they observed, is complex, and "The only successful account of the origin of complex biological structure is the theory of natural selection, the view that the differential reproductive success associated with heritable variation is the primary organizing force in the evolution of organisms."[5] They go on to point out that the emergence of complex structure through natural selection is gradual: "The only way for complex design to evolve is through a sequence of mutations with small effects."[6] On *a priori* grounds, then, it seems highly unlikely that language was the product of a single mutation in some lone Prometheus.

As we saw in the previous chapter, there is still general agreement that the critical ingredient distinguishing most human languages from other forms of animal communication is recursive grammar. Even this ingredient need not have appeared suddenly and fully formed in our own species, and indeed, as we saw in chapter 2, there may be languages such as that of the Pirahã that do not make use of recursion. The concept of grammaticalization, also discussed in chapter 2, suggests scenarios in which grammar unfolds gradually, rather than in a single step. Instead of supposing that it all happened within the past 100,000 years, it seems much more reasonable to suppose that grammatical language evolved slowly—and variably—over the six or seven million years since the hominins parted company with the chimpanzee line, although I suggest below that it was probably the last two million years that were critical. This must still be regarded as something of a big bang in evolutionary terms, but does give the theorist more breathing space to develop a plausible evolutionary scenario. But of course at least some of the other ingredients of language were no doubt present in our primate forebears, and to understand how language evolved we need to reach back in time to these forebears, then forward to hominin evolution, and then try to determine what gave language its power of limitless expression.

In the previous chapter, I showed that the nearest equivalents of language in nonhuman primates lie in manual systems rather than in vocal calls. Manual activity in primates is intentional and subject to learning, whereas vocalizations appear to be largely involuntary and fixed.[7] In teaching great apes to speak, much greater success has been achieved through gesture and the use of keyboards than through vocalization, and the bodily gestures of apes in the wild are less constrained by context than are their vocalizations. These observations strongly suggest that language evolved from manual gestures.

The Gestural Origins of Language

The theory that language evolved from manual gestures has a long but checkered history. An early advocate was the eighteenth-century French philosopher Abbé Étienne Bonnot de Condillac. He was interested in how language evolved, but as a priest was on dangerous ground, since the theological view was that language was a gift from God. In order to express his own heretical view he therefore had to present it as a fable.[8] He imagined two abandoned children, a boy and a girl, who had not yet acquired language and were wandering about in the desert after the Flood. In order to communicate they used manual gestures. If the boy wanted something out of his reach, "He did not confine himself to cries or sounds only; he used some endeavors to obtain it, he moved his head, his arms, and every part of his body." These movements were understood by his companion, who was then able to help. Eventually there grew "a language which in its infancy, probably consisted only in contortions and violent agitations, being thus proportioned to the slender capacity of this young couple."[9]

The story goes on to explain how articulated sounds came to be associated with gestures, but "the organ of speech was so inflexible that it could not articulate any other than a few simple sounds."[10] Eventually, though, the capacity to vocalize increased, and "appeared as convenient as the mode of speaking by action; they were both indiscriminately used; till at length articulate sounds became so easy, that they absolutely prevailed."[11]

On the surface, at least, this story is not about the evolution of language, but is rather about how two stranded children developed a way of communicating. It is likely, though, that Condillac really intended it to be the story of how language evolved as a human faculty, and it was remarkably prescient.

The idea that language evolved from manual gestures has since been proposed many times, although not always accepted. Condillac's near contemporary, Jean-Jacques Rousseau, evidently unfazed by religious prohibitions, endorsed the gestural theory more openly in an essay published in 1782. Charles Darwin at least pointed (as it were) to it: "I cannot doubt that language owes its origins to the imitation and modification of various natural sounds, and man's own distinctive cries, *aided by signs and gestures*."[12] In 1900, Wilhelm Wundt, the founder of the first laboratory of experimental psychology at Leipzig in 1879, wrote a two-volume work on speech, and argued that a universal sign language was the origin of all languages.[13] He wrote, though, under the misapprehension that all deaf communities use the same system of signing, and that signed languages are useful only for basic communication, and are incapable of communicating abstract ideas. We now know that signed languages vary widely from community to community, and can have all of the communicative sophistication of speech.

The British neurologist MacDonald Critchley lamented that his book *The Language of Gesture* coincided with the outbreak of World War II and was therefore largely ignored, so he wrote a second book called *Silent Language*, which was published in 1975. "Gesture," he wrote, "is full of eloquence to the sagacious and vigilant onlooker who, holding the key to its interpretation, knows how and what to observe."[14] Critchley was ambivalent about whether language originated in gesture, being at one point unable to accept that language could have been at one time gestural and voiceless, but later arguing that gesture must have predated speech in human evolution.

Perhaps the first comprehensive case for the gestural theory of language origins in modern times was presented by the anthropologist Gordon W. Hewes in an article published in 1973. Hewes was partly motivated by the discovery that great apes could not be taught to speak, but were reasonably successful at using signs to communicate. The gestural theory was strengthened by the work of Ursula Bellugi and Edward S. Klima, revealing American Sign Language (ASL) to be a full language, affected by specific brain injury in very much the same way that spoken language is.[15] The gestural theory then seemed to lie dormant for a while. I picked it up in my 1991 book *The Lopsided Ape*, which was shortly followed by the 1994 book *Gesture and the Nature of Language*, by William C. Stokoe, David F. Armstrong, and Sherman E. Wilcox, who approached it from the perspective of signed languages. This was followed by Armstrong's book *Original Signs* in 1999. I elaborated my own views in my 2002 book, *From Hand to Mouth*, and in the previous year William C. Stokoe published *Language in Hand*, with the self-explanatory subtitle *Why Sign Came before Speech*.[16]

The gestural theory received a powerful boost, though, with the remarkable discovery of so-called mirror neurons in the primate brain.

Mirror Neurons

In 2000, the neuroscientist Vilayanur Ramachandran famously remarked that mirror neurons would do for psychology what DNA has done for biology[17]—a remark that is in danger of being quoted almost as often as mirror neurons themselves are invoked. Mirror neurons were discovered in the monkey brain by the Italian scientist Giacomo Rizzolatti and his colleagues at the University of Parma. The activity of these neurons was recorded from electrodes inserted into a part of the frontal cortex called area F5. They form a subset of a class of neurons that are active when the monkey makes an intentional movement of the hand, such as reaching to grasp an object, like a peanut. To Rizzolatti's initial surprise, some of these neurons also responded when the monkey observed another individual (such as the researcher) making the same movement. These are the neurons dubbed "mirror neurons," because perception is mirrored in action. They have also been called "monkey see, monkey do" neurons.

The idea that mirror neurons may have set the stage for the eventual evolution of language was first set out by Michael Arbib and Giacomo Rizzolatti.[18] The main points are as follows.

First, area F5 is homologous to an area in the human brain known as Broca's area, which plays a critical role in speech and language. More precisely, Broca's area can be divided into two areas, known as Brodman areas 44 and 45, and area 44 is considered the true analogue of area F5. In humans, it is now evident that area 44 is involved not only in speech, but also in motor functions unrelated to speech, including complex hand movements, and sensorimotor learning and integration.[19] In the course of human evolution, then, it seems that vocalization must have been incorporated into the system, which explains why language can be either vocal, as in speech, or manual, as in signed languages.[20]

Second, mirror neurons are now understood to be part of a larger network, called the mirror *system*. In the monkey it includes, besides area F5, more posterior areas such as the superior temporal sulcus and the inferior parietal lobule.[21] This system largely over-

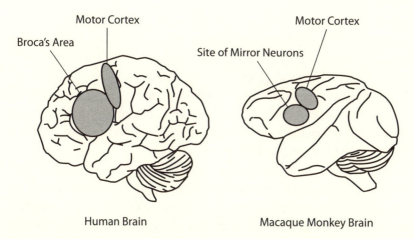

Figure 7. Sites of Broca's area and motor cortex in the human brain (*left*), and of mirror neurons and motor cortex in the macaque brain (*right*). Copyright © 2010 W. Tecumseh Fitch. Reprinted with the permission of Cambridge University Press.

laps the corresponding regions in the human brain that have to do with the more general functions of language. Besides Broca's area, these regions include the other well-known language area, Wernicke's area, in the posterior part of the superior temporal sulcus, although language areas are probably distributed more widely than these two classic areas.[22] The overlap has led to the notion that language grew out of the mirror system itself, an idea developed in some detail by Michael Arbib.[23]

Third, Rizzolatti and colleagues proposed that the mirror system in the monkey is in essence a system for understanding action. That is, the monkey understands the actions of others in terms of how it would itself perform those actions. This is the basic idea underlying what has been called the *motor theory of speech perception*, which holds that we perceive speech, not in terms of the acoustic patterns it creates, but in terms of how we ourselves would articulate it. This theory arose from the work the Alvin Liberman and others at the Haskins Laboratories in the United States, who sought the acoustic principles underlying the basic units of sound that make up our speech.[24] For example, *b* sounds in words like *battle, bottle, beer, bug, rabbit, Beelzebub,* or *flibbertigibbet* probably sound much the same to you, but the actual acoustic streams created by these *b* sounds varies widely, to the point that they actually have virtually nothing in common.[25] The same is true of other speech sounds, especially the plosive sounds *d*, *g*, *p*, *t*, and *k*; the acoustic signals vary widely depending on the contexts in which they are embedded. Liberman and colleagues concluded that we hear each sound as the same in each case because we "hear" it in terms of how we produce it.

Fourth, we saw in the previous chapter that vocalization in primates seems to be largely involuntary and, for the most part at least, impervious to learning. The mirror system, in contrast to the primate vocalization system, has to do with intentional action, and is clearly modifiable through experience. For example, mirror neurons in the monkey brain respond to the *sounds* of certain actions, such as the tearing of paper or the cracking of nuts,[26] and these responses can only have been learned. The neurons were not activated, though, by monkey calls, suggesting that vocalization itself is not part of the mirror system in monkeys. In our forebears, then, the mirror system was already set up for processing the sounds caused by manual activity, but not for the processing of vocal sounds.

Of course, nonhuman primates do not have language as we know it, but the mirror system provided a natural platform for language to evolve. Indeed the mirror system is now well documented in humans, and involves characteristics that are more language-like than those in the monkey. For example, in the monkey, mirror neurons respond to transitive acts, as in reaching for an actual object, but do not respond to intransitive acts, where a movement is mimed and involves no object.[27] In humans, in contrast, the mirror system responds to both transitive and intransitive acts, and the incorporation of intransitive acts would have paved the way to the understanding of acts that are symbolic rather than object-related.[28] More directly, though, functional magnetic resonance imaging (fMRI) in humans shows that the mirror-neuron region of the premotor cortex is activated not only when they watch movements of the foot, hand, and mouth, but also when they read phrases pertaining to these movements.[29] Somewhere along the line, the mirror system became interested in language.

An Evolutionary Scenario

Let us suppose, then, that intentional communication grew out of action understanding in our primate forebears. In the previous chapter, we saw that a number of great apes have learned language-like gestures, suggesting that the incorporation of intransitive gesture had already occurred in our great-ape forebears. These gestures lack grammar, and are therefore not true language, but it is reasonable to suppose that similar activity was a precursor to language in our hominin forebears who separated from the line leading to chimpanzees and bonobos some six or seven million years ago.

Unlike their great-ape cousins, the hominins were bipedal, which would have freed the hands for the further development of expressive manual communication. The body and hands are free to move in four dimensions (three of space and one of time), and so mimic activity in the external world. The hands can also assume, at least approximately, the shapes of objects or animals, and the fingers can mimic the movement of legs and arms. The movements of the hands can also mimic the movement of objects through space, and facial expressions can convey something of the emotions of events being described. Mimesis persists in dance, ballet, and mime, and we all resort to mime when trying to communicate with people who speak a language different from our own. Once, in Russia, I was able to successfully request a bottle opener by miming the action of opening a beer bottle, to the vast amusement of the people at the hotel desk.

Although predominantly bipedal, the early hominins were still partially adapted to arboreal life, and walked only clumsily on two legs. This is known as facultative bipedalism. Merlin Donald, in his 1991 book *Origins of the Human Mind*, suggested that what he called "mimetic culture" did not evolve until the emergence of *Homo ergaster* from around two million years ago. In *Ergaster* and the later members of the genus *Homo*, moreover, bipedalism shifted from facultative to obligate—that is, it became obligatory, and assumed a more free-striding gait. More critically, perhaps, brain size began to increase dramatically with the emergence of the genus *Homo*, which might be taken as evidence of selection for more complex communication, and perhaps the beginnings of grammar.

Even in modern humans, mimed action activates the brain circuits normally thought of as dedicated to language. In one experiment, for example, brain activity was recorded while people attended to video clips of a person performing pantomimes of actions, such as threading a needle, or what are called *emblems*, such as lifting a finger to the lips to indicate quiet. Activity was also recorded while the subjects gave spoken descriptions of these actions. All three activities elicited activity in the left side of the brain in frontal and posterior areas—including Broca's and Wernicke's areas—that have been identified since the nineteenth century as the core of the language system. The authors of this study conclude that these areas have to do, not just with language, but with the more general linking of symbols to meaning, whether the symbols are words, gestures, images, sounds, or objects.[30]

We also know that the use of signed language in the profoundly deaf activates the same brain areas that are activated by speech,[31] and indeed modern sign languages are also partly dependent on mime. It has been estimated, for example, that in Italian Sign Language some 50 percent of the hand signs and 67 percent of the bodily locations of signs stem from iconic representations, in which there is a degree of spatiotemporal mapping between the sign and its meaning.[32] In American Sign Language, too, some signs are arbitrary, but many more are iconic. For example, the sign for *erase* resembles the action of erasing a blackboard, and the sign for *play piano* mimics the action of actually playing a piano.[33] But of course signs need not be transparently iconic, and the meanings of even iconic symbols often cannot be guessed by naïve observers, or even by those using a different sign language.[34] They also tend to become less iconic and more arbitrary over time, in the interests of speed, efficiency, and grammatical constraints. This process is known as *conventionalization*.[35]

The Swiss linguist Ferdinand de Saussure wrote of the "arbitrariness of the sign" as a defining property of language,[36] and on this basis it is sometimes supposed that signed languages, with their

American	Chinese	Danish
sign language	sign language	sign language

Figure 8. Signs for TREE in different sign languages. Although all are fundamentally iconic, they differ markedly. Iconic representations are not always transparent to the viewer not conversant with the language (author's drawing).

strong basis in iconic representations, are not true languages. Although most words in spoken languages are indeed arbitrary—the words *cat* and *dog* in no way resemble those friendly animals or the sounds that they make—there are of course some words that are onomatopoeic. One such word is *zanzara*, which is the evocative Italian word for mosquito, and Steven Pinker notes a number of newly minted examples: *oink, tinkle, barf, conk, woofer, tweeter.*[37] We should perhaps add *twitter*. Speech can also mimic visual properties in subtle ways; for example, it has been shown that speakers tend to raise the pitch of their voice when describing an object moving upwards, and lower it in describing a downward movement.[38] The arbitrariness of words (or morphemes) is not so much a necessary property of language, though, as a matter of expedience, and of the constraints imposed by the particular language medium.

Speech, for example, requires that the information be *linearized*, piped into a sequence of sounds that are necessarily limited in terms of how they can capture the spatial and physical natures of what they represent. The linguist Charles Hockett put it this way:

When a representation of some four-dimensional hunk of life has to be compressed into the single dimension of speech, most iconicity is necessarily squeezed out. In one-dimensional projections, an elephant is indistinguishable from a woodshed. Speech perforce is largely arbitrary; if we speakers take pride in that, it is because in 50,000 years or so of talking we have learned to make a virtue of necessity.[39]

Signed languages are clearly less constrained. The hands and arms can mimic the shapes of real-world objects and actions, and to some extent lexical information can be delivered in parallel instead of being forced into rigid temporal sequence. With the hands, it is almost certainly possible to distinguish an elephant from a woodshed, in purely visual terms. Even so, conventionalization allows signs to be simplified and speeded up, to the point that many of them lose most or all of their iconic aspect. For example, in American Sign Language the sign for *home* was once a combination of the sign for *eat*, which is a bunched hand touching the mouth, and the sign for *sleep*, which is a flat hand on the cheek. Now it consists of two quick touches on the cheek, both with a bunched handshape, so the original iconic components are effectively lost.[40]

Although signing has been recorded from as early as 360 BC,[41] modern signed languages have short pedigrees, arising independently among different deaf communities. This somewhat contaminates the comparison between signed and spoken languages, which have evolved, albeit with modification and divergence, over tens of thousands of years. One interesting exception is Turkish Sign Language, which has highly schematized morphology and an exceptionally large proportion of arbitrary, noniconic signs.[42] Turkish Sign Language may go back over 500 years. Visitors to the Ottoman court in the sixteenth century observed that mute servants, most of them deaf, were favored in the court, probably because they could not be bribed for court secrets. These servants developed a sign language, which was also acquired by many of the courtiers. A photograph published in 1917 shows two servants still using sign language. It is not known for sure whether modern Turkish Sign Language is related to that of the Ottoman court, but if it is, it supports the view that the passage of time is the critical element in the loss of iconic representation.[43]

Language, then, may have evolved from mime, with the arbitrary nature of some words or signs deriving from the drive to greater economy and from the constraints of the medium through which language is expressed. Conventionalization of course depends on the power of the brain to form associations, since the iconic or onomatopoeic component that may serve to indicate meaning is effectively lost. We know from the studies of Kanzi and the other "linguistic apes" that the ability to form such associations is not unique to humans, although the human capacity to do so may far outstretch that of the ape. Kanzi has learned a few hundred symbols, but the average literate human has a vocabulary on the order of 50,000 words, most of them neither iconic nor onomatopoeic.[44] This may be one reason why the human brain is some three times larger relative to body size than that of the great apes.[45] We simply need a much larger dictionary.

Once conventionalization sets in, there is no reason why language need be restricted to the visual domain. Spoken words will do as well as signed ones. But this of course raises the question of why language switched, at least in the majority of humans, from manual gesture to speech.

The Switch

This indeed has proven to be the most contentious issue for the gestural theory. The linguist Robbins Burling, for example, writes

> [T]he gestural theory has one nearly fatal flaw. Its sticking point has always been the switch that would have been needed to move from a visual language to an audible one.[46]

In another recent book, Peter F. MacNeilage expresses similar concerns,[47] but I will argue that the switch was probably a relatively simple and natural one.

The first step from manual gesture to speech may have been the incorporation of facial gestures. Even in the monkey, manual and facial gestures are closely linked neurophysiologically, as well as behaviorally.[48] Some neurons in area F5 fire when the animal makes movements to grasp an object with either the hand or the mouth. An area in the monkey brain that is considered homologous to Broca's area is involved in control of the orofacial musculature—though not of speech itself.[49] These neural links between hand and mouth may be related to eating rather than communication, perhaps involved in preparing the mouth to hold an object after the hand has grasped it, but later adapted for gestural and finally vocal language.

The connection between hand and mouth can also be demonstrated in human behavior. In one experiment, people were instructed to open their mouths while grasping objects with their hands, and the size of the mouth opening increased with the size of the grasped object. Conversely, when they open their hands while grasping objects with their mouths, the size of the hand opening also increased with the size of the object.[50] Grasping movements of the hand also affect the way we utter sounds. If people are asked to say *ba* while grasping an object, or even while watching someone else grasping an object, the syllable itself is affected by the size of the object grasped. The larger the object, the wider the opening of the mouth, with consequent effects on the speech formants.[51] These effects can also be observed in one-year-old infants.[52]

The link between hand and mouth suggests that early communicative gestures may have involved facial gestures as well as gestures of the hands. Indeed, speech itself retains a strong visual component. This is illustrated by an effect attributed to the psychologist Harry McGurk,[53] in which dubbing sounds onto a mouth that is saying something different alters what the hearer actually hears; that is, the viewer/listener often reports what the speaker is seen to be saying rather than the speech sound itself, or sometimes a blend of the two. Deaf people often become highly proficient at lipreading, and even in individuals with normal hearing the brain areas involved in speech production are activated when they view speech-related lip movements.[54] Ventriloquists project their own voices onto the face of a dummy by synchronizing the mouth movements of the dummy with their own pursed-lipped utterances.

The signed languages of the deaf also depend at least as much on movements of the face as of the hands. Facial expressions and head movements can alter the mood or polarity of a sentence. For

example, a question in American Sign Language is signaled by raising the eyebrows, and shaking the head while signing a sentence
turns it from affirmative to negative. Movements of the mouth are
especially important, to the point that linguists are beginning to
identify the rules that govern the formation of mouthed signs.[55]
Mouth gestures can also serve to disambiguate hand gestures, and
as part of more general facial gestures provide the visual equivalent of prosody in speech.[56] Recordings of eye movements show
that users of British Sign Language watch the face more often than
they watch the hands or body when watching a person signing a
story.[57]

The first part of the switch, then, was probably the increasing involvement of the face. The idea that movements of the face played
a role in the evolution of language was anticipated by Friedrich
Nietzsche in his 1878 book *Human, All Too Human*. Aphorism
216 from that book reads in part as follows;

> Imitation of gesture is older than language, and goes on involuntarily
> even now, when the language of gesture is universally suppressed, and
> the educated are taught to control their muscles. The imitation of ges-
> ture is so strong that we cannot watch a face in movement without the
> innervation of our own face (one can observe that feigned yawning
> will evoke natural yawning in the man who observes it). The imitated
> gesture led the imitator back to the sensation expressed by the gesture
> in the body or face of the one being imitated. This is how we learned to
> understand one another; this is how the child still learns to understand
> its mother. In general, painful sensations were probably also expressed
> by a gesture that in its turn caused pain (for example, tearing the hair,
> beating the breast, violent distortion and tensing of the facial muscles).
> Conversely, gestures of pleasure were themselves pleasurable and were
> therefore easily suited to the communication of understanding (laugh-
> ing as a sign of being tickled, which is pleasurable, then served to ex-
> press other pleasurable sensations).
>
> As soon as men understood each other in gesture, a symbolism of
> gesture could evolve. I mean, one could agree on a language of tonal
> signs, in such a way that at first both tone and gesture (which were
> joined by tone symbolically) were produced, and later only the tone.

This remarkable extract also anticipates the discovery of the mirror system.

The final act, then, was the incorporation of vocalization. Part of the reason for this may have been that facial gestures increasingly involved movements of the tongue that are invisible, and activation of the vocal cords simply allowed these invisible gestures to be accessible. Speech might be described as facial gesture half swallowed, with sound added. Along with the incorporation of vocal sound, the vocal tract itself changed, and control of the tongue was enhanced, enabling a wider diversity of sounds. The sound itself could be turned on or off to create the distinction between voiced sounds, such as the consonants /b/, /d/, and /g/, and their unvoiced equivalents, /p/, /t/, and /k/.

Speech itself can be viewed as gestural rather than acoustic.[58] The motor theory of speech perception, described above, is based on the idea that perceiving speech sounds depends on the mapping of those sounds on to the articulatory gestures that produce them. This has led to what is known as articulatory phonology,[59] in which speech is understood as gestures produced by six articulatory organs, the lips, the velum, the larynx, and the blade, body, and root of the tongue. In the context of speech understood as gesture, then, the incorporation of vocal gestures into the mirror system may have been a relatively small step for mankind.

But how and when did it happen?

FOXP2

A possible clue comes from genetics. About half of the members of three generations of an extended family in England, known as the KE family, are affected by a disorder of speech and language. The disorder is evident from the affected child's first attempts to speak and persists into adulthood.[60] The disorder is now known to be due to a point mutation on the FOXP2 gene (forkhead box P2) on chromosome 7.[61] For normal speech to be acquired, you need two functional copies of this gene.

The FOXP2 gene, then, may have played a role in incorporating vocalization into the mirror system, thus allowing speech to develop as an intentional, learnable system.[62] As we saw earlier, one of the main brain centers for the production of speech is Broca's area, which is also part of the mirror system. Typically, Broca's area is activated when people generate words. A brain-imaging study revealed, though, that the members of the KE family affected by the mutation, unlike their unaffected relatives, showed no activation in Broca's area while silently generating verbs.[63] The FOXP2 gene may also play a role in other species. In songbirds, knockdown of the FOXP2 gene impairs the imitation of song,[64] and insertion of the FOXP2 point mutation found in the KE family into the mouse critically alters synaptic plasticity and motor learning.[65] On the other hand, insertion of the normal human variant of FOXP2 into the mouse gives rise to qualitatively different ultrasonic variations.[66] The role of FOXP2 in foxes is unknown.

FOXP2 is highly conserved in mammals, and in humans the gene differs in only three places from that in the mouse. Nevertheless it underwent two mutations since the split between hominin and chimpanzee lines. According to one theoretical estimate, the more recent of these occurred "not more than" 100,000 years ago,[67] although the error associated with this estimate makes it not unreasonable to suppose that it coincided with the emergence of Homo sapiens around 170,000 years ago. This conclusion is seemingly contradicted, though, by recent evidence that the mutation is also present in the DNA of a 45,000-year-old Neandertal fossil,[68] suggesting that it goes back at least 700,000 years to the common ancestor of humans and Neandertals.[69]

But this is challenged in turn by a more recent phylogenetic dating of the haplotype, in which the time of the most recent common ancestor carrying the FOXP2 mutation was estimated at 42,000 years ago, with 95 percent confidence that it occurred between 38,000 and 45,500 years ago.[70] Even allowing for distortions in assumptions underlying this estimate, it is much more consistent with the earlier molecular-based estimate than with that based on the Neandertal fossil. It is possible that the Neandertal DNA was contaminated, or that there was some degree of inbreeding between Neandertal and Homo sapiens, who overlapped in Europe for some 20,000 years. Recent evidence suggests that microcephalin, a gene involved in regulating brain size, may have entered the human gene pool through interbreeding with Neandertals,[71] so the reverse possibility of *FOXP2* entering the late Neandertal gene pool from *Homo sapiens* is not completely ruled out. We might have been slightly chummier with the Neandertals than is generally thought.[72]

It may well be that the *FOXP2* gene holds the secret of the emergence of speech. It may nevertheless be a chimera, since it is known to be involved in a wide range of areas, not only in the brain, but also in the heart, lung, and gut.[73] Moreover, the mutated gene in the KE family is not the same as that in the last common ancestor of chimpanzee and human. Even so, it continues to attract interest as a likely clue to how speech evolved.[74]

Anatomical Changes

Be that as it may, there is other evidence that the final touches making autonomous speech possible may not have been complete in the Neandertal. One requirement for articulate speech was the lowering of the larynx, creating a right-angled vocal tract that allows us to produce the wide range of vowels that characterize speech. Philip Lieberman has argued that this modification was incomplete even in the Neandertals, a species of *Homo* with brains as large as those of *Homo sapiens*[75]—slightly larger, in fact, but let's not go there. We share a common ancestry with the Neandertals going back some 700,000 years, but parted company some 370,000 years ago before again sharing territory in Europe from around 50,000 years ago.[76] The Neandertals were driven to extinction around 30,000 years ago, suggesting that we humans may have talked them out of existence—perhaps a more palatable notion, as it were, than that we simply slaughtered them.

Lieberman's views on this are controversial.[77] In direct contradiction, Tattersall writes that "a vocal tract had . . . been achieved among humans well over half a million years before we have any independent evidence that our forebears were using language or speaking."[78] The emergence of language, in Tattersall's view as in Chomsky's, was thus a later event, unrelated to the capacity for speech. Lieberman has nevertheless received support from his son

Daniel Lieberman, who has shown that the structure of the cranium underwent changes after we split with the Neandertals. One such change is the shortening of the sphenoid, the central bone of the cranial base from which the face grows forward, resulting in a flattened face.[79] This flattening may have been part of the change that created the right-angled vocal tract, with horizontal and vertical components of equal length.[80] This is the modification that allowed us the full range of vowel sounds, from *ah* to *oo*. Another adaptation unique to *H. sapiens* is neurocranial globularity, defined as the roundness of the cranial vault in the sagittal, coronal, and transverse planes,[81] which is likely to have increased the relative size of the temporal and frontal lobes relative to other parts of the brain.

Other anatomical evidence suggests that the anatomical requirements for fully articulate speech were probably not complete until late in the evolution of *Homo*. For example, the hypoglossal canal is much larger in humans than in great apes, suggesting that the hypoglossal nerve, which innervates the tongue, is also much larger in humans, perhaps reflecting the importance of tongued gestures in speech. The evidence suggests that the size of the hypoglossal canal in early australopithecines, and perhaps in *Homo habilis*, was within the range of that in modern great apes, while that of the Neandertal and early *H. sapiens* skulls was contained well within the modern human range,[82] although this has been disputed.[83] A further clue comes from the finding that the thoracic region of the spinal cord is relatively larger in humans than in nonhuman primates, probably because breathing during speech involves extra muscles of the thorax and abdomen. Fossil evidence indicates that this enlargement was not present in the early hominins or even in *Homo ergaster*, dating from about 1.6 million years ago, but was present in several Neandertal fossils.[84]

Emboldened by such evidence, and no doubt heartened by familial support, Philip Lieberman has recently made the radical claim that "fully human speech anatomy first appears in the fossil record in the Upper Paleolithic (about 50,000 years ago) and is absent in both Neandertals and earlier humans."[85] This provocative statement suggests that articulate speech emerged even later than the arrival of *Homo sapiens* some 150,000 to 200,000 years ago. While this may be an extreme conclusion, the bulk of evidence does suggest that autonomous speech emerged very late in the human repertoire. Perhaps the critical question is whether the capacity to speak was unique to humans, or whether we shared it with the Neandertals. I return to this question in chapter 12.

Why the Switch?

It has become clear that the signed languages of the deaf have all of the linguistic sophistication of spoken languages. At Gallaudet University, in Washington, DC, instruction is entirely in American Sign Language, and includes all the usual academic disciplines, even poetry. In some respects, signed languages may have an advantage, since the greater iconic component can provide extra clues to meaning.

This aside, then, the advantages of speech over manual language are likely to be practical rather than linguistic. Let's consider what these advantages might be.

Spatial Reach

Sound reaches into areas inaccessible to sight. You can talk to a person who is hidden from sight, whereas signed language requires visual contact. This has the important advantage of allowing communication at night, especially during the times when there was no artificial lighting, except perhaps for the campfire. The San, a modern hunter-gatherer society, are known to talk late at night, sometimes all through the night, to resolve conflict and share knowledge.[86] Mary Kingsley, the noted British explorer of the late nineteenth century, made the following observation of tribes she encountered in Africa:

> African languages [are not elaborate enough] to enable a native to state his exact thought. Some of them are very dependent upon gesture. When I was with the Fans they frequently said "We will go to the fire so that we can see what they say," when any question had to be decided after dark, and the inhabitants of Fernando Po, the Bubis, are quite unable

to converse with each other unless they have sufficient light to see the accompanying gestures of the conversation.[87]

While this may seem condescending, it may well be the case that some cultures make more use of gesture than others,[88] through cultural rather than biological necessity.

Vocal language can carry over longer distances than signed language, but its reach is not simply one of distance. You can talk to someone in the next room, or even while the person's eyes are closed. Many an intelligent question after a talk has come from a member (usually elderly) of the audience who has appeared to be asleep throughout the talk. Signed language requires that the recipient is actually looking at the signer, whereas speech enters the ears regardless of how the listener is oriented or placed relative to the speaker. We speak of the *line* of sight, but the *envelope* of sound—and a three-dimensional envelope at that. Speech works over longer distances than signing, and can more readily command attention. You can wake a sleeping audience by shouting, but no amount of silent gesture will do the trick.

Speech gains these physical advantages with little cost compared to gesture. Teachers of sign language have been known to report needing regular massages in order to meet the sheer physical demands of sign-language expression. In contrast, the physiological costs of speech are so low as to be nearly unmeasurable.[89] In terms of expenditure of energy, speech adds little to the cost of breathing, which we must do anyway to sustain life.

Diversity and the Language Fortress

In chapter 2, I noted that there is huge diversity in the world's languages, allowing language to serve as a badge of a culture's distinctiveness. There may also have been pressure to establish languages that were impenetrable to outsiders, thereby enhancing group membership and keeping out intruders and freeloaders. The absence of any iconic component in sound-based languages reduces penetrability, and the sheer diversity of possible sound-based systems adds further to the fortress-like nature of language. Although there are diverse sign languages, the diversity is almost certainly tiny compared with that possible in the deployment of speech sounds, even allowing for the fact that existing signed languages have a much shorter pedigree.

The impenetrability of speech is well illustrated by an anecdote from World War II. The outcome of the war in the Pacific hinged to a large extent on the ability of each side to crack codes developed by the other side. Early in the war, the Japanese were easily able to crack the codes developed by the allies, but the American military then developed a code that proved essentially impenetrable. They simply employed Navajo speakers to communicate openly in their own language via walkie-talkie. To the Japanese, Navajo simply sounded like a "strange, gurgling" sound, unrecognizable even as language.[90]

Nicholas Evans asserts that there are well over 1,500 possible speech sounds, but no language uses more than 10 percent of them for its inventory of phonemes. As we have seen, phonemes do not map exactly onto actual sounds, but some measure of the variation in sounds used can be gained by comparing the number of phonemes identified for different languages. The most parsimonious of languages appears to be that spoken by women of the Pirahã, the small hunter-gatherer Amazonian tribe in Brazil that we encountered in chapter 2. The Pirahã language as spoken by women has only seven consonants and three vowels.[91] As though to compensate for a male lack of verbal fluency, the men are permitted eight consonants and three vowels; with 11 phonemes, they tie with Hawaiian for the second most parsimonious language.[92] New Zealand Maori has only 14 phonemes, but Maori are known for fine oratory. In Maori society, as in other traditionally oral societies, speech implies power and status; the New Zealand scholar Anne Salmond writes that, among the Maori, "oratory is the prime qualification for entry into the power game."[93] The most phonologically diverse language may be !Xóõ, spoken by about 4,000 people in Botswana and Namibia, which has somewhere between 84 and 159 consonants.[94] English bumbles along with about 44 phonemes.

Different languages, then, can in principle use inventories of sounds that are totally distinct, or very nearly so. Evans gives the example of the ǀGui language of Botswana, which includes a large number of click sounds unpronounceable to those who don't speak the language. Conversely, a Japanese colleague of Evans found the ǀGui quite unable to pronounce her own name, which is *Mimmi*—straightforward enough to speakers of Western languages. Even within Western languages, there are sounds in one language that speakers of other languages cannot produce. Japanese people have difficulty distinguishing between *l*- and *r*-sounds, creating difficulties with words like *parallel* or Dylan Thomas's *Llareggub*.[95] Speakers of English have comparable difficult distinguishing the different *t*-sounds of Hindi, and I never could quite get the precise form of the French *r*-sound—*aarrghh*!

Paradoxically, none of the sounds used in speech poses a problem to children exposed from an early age. There is no reason to suppose a child born to a ǀGui family but raised from birth in New York would have any difficulty learning to speak New York English, which to me sounds only slightly peculiar, and would no doubt eventually lose any ability to learn ǀGui. On the other hand, Woody Allen would have easily learned ǀGui had he been exposed to it from infancy. It has been claimed that very young infants can discriminate between many, if not all, of the phonemes of the world's languages,[96] but by the age of about one they can discriminate only the phonemes of the language or languages they have been exposed to.

Little babies have the potential to crack the language barrier erected by different groups, but by the time we are adults, language makes outsiders of us all.

Freeing the Hands

Charles Darwin wrote: "We might have used our fingers as efficient instruments [of communication], for a person with practice can report to a deaf man every word of a speech rapidly delivered at a public meeting; but the loss of our hands, while thus employed, would have been a serious inconvenience."[97] Darwin made this point to account for why we evolved speech rather than signed language, but the argument holds equally as an explanation for a switch from manual gesture to speech.

The switch, then, would have freed the hands for other activities, such as carrying and manufacture. It also allows people to speak and use tools at the same time. It might be regarded, in fact, as an early example of miniaturization, whereby gestures are squeezed from the upper body to the mouth. It also allows the development of pedagogy, enabling us to explain skilled actions while at the same time demonstrating them, as in a modern television cooking show.[98] The freeing of the hands and the parallel use of speech may have led to significant advances in technology,[99] and help explain why humans eventually predominated over other large-brained hominins, including the Neandertals, who died out some 30,000 years ago. I take up this theme in more detail in chapter 12, but suffice to say here that our newfound manual freedom may have been primarily responsible for the evolution of what anthropologists call "modernity," which makes us so markedly different from all of our hominin cousins.

And Still We Gesture

Of course there are also advantages to visual language over vocal language. Vocal language is denied to those unable to hear or to speak, and signed languages form a natural substitute. Visual language still permits an iconic element, and most people resort to gesture, or drawing, when trying to communicate with those who speak a different language, or even when trying to explain spatial concepts, such as a spiral. Some sort of manual gesture is necessary even for the acquisition of speech; in learning the names of objects, for example, there must be some means of indicating which object has which name. Even adults gesture as they speak, and their gestures can add a significant component of meaning.[100] For example, people regularly point to indicate directions: *He went that way . . .*, accompanied by pointing. Language has still not fully escaped its manual origins. Indeed, speech could scarcely exist without gesture, some way to relate words to the physical world. As Ludwig Wittgenstein put it, "It isn't the colour red that takes the place of the word red, but the gesture that points to a red object."[101]

And of course what people do is often as eloquent as the things they say. In Shakespeare's *Henry VIII*, Norfolk says this of Cardinal Wolsey:

Some strange commotion
Is in his brain; he bites his lip and starts;
Stops on a sudden, looks upon the ground,
Then, lays his finger on his temple; Straight,
Springs out into a fast gait; then, stops again,
Strikes his breast hard; and anon, he casts
His eye against the moon: In most strange postures
We have seen him set himself.

In this chapter, we have come a long way from Prometheus, that solitary pioneer who suddenly found himself possessed of language. Even so, we have seen some evidence of late changes, perhaps restricted to *Homo sapiens*, that gave rise to the power of speech. This may even have involved mutations, such as that of the *FOXP2* gene. But this does not mean of course that *language* emerged late. Along the way, though, as we progressed from manual gesture to speech, there is still one missing ingredient—grammar. How and when did language, whether spoken or signed, acquire its distinctive property of generativity, the power to create a potentially infinite number of different sentences? For an answer to that, we need to digress first through two other recursive capacities arguably unique to humans, episodic memory and mental time travel. We shall then find one likely source of grammatical language in the human ability to transcend time.

―

PART 2

Mental Time Travel

We humans can effortlessly transport ourselves mentally to other places and other times. We remember specific events, imagine possible future ones, and even invent fictional ones. Experimental psychologists have typically studied this power in the context of memory, asking people to imagine earlier events in their lives, or providing information and then asking people to recall it. More recently, though, the focus has shifted to the imagining of future events, revealing a continuity between so-called episodic memory and what has been termed episodic foresight. In both cases, there is a constructive element—even memory for specific episodes deviates substantially from what actually happened. Because of the constructive nature of both episodic foresight and episodic memory, mental time travel blends naturally into fiction.

Chapter 5 begins with memory, and especially so-called episodic memory, which involves the conscious reliving of past episodes. In chapter 6 I extend this focus to the more general notion of mental time travel, which includes the imagining of possible future episodes. In chapter 7 I develop the theme that mental time travel is critical to language. Indeed one of the so-called design features of language, as specified by the American linguist Charles Hockett, is *displacement*—the capacity to refer to events removed from the here and now.[1] Language may therefore have evolved primarily to enable humans to share their memories, plans, and stories, enhancing social cohesion and creating a common culture. At the same time, the very mechanisms that create cohesion within groups can instill antagonisms between them—as the American writer and civil rights activist James Arthur Baldwin once put it, "People are trapped in history, and history is trapped in them."

This section heralds a departure from the conventional view of language as an independent biological capacity, and suggests instead that language is deeply bound to culture. The "language fortress" describe in chapter 4 is part of the cultural fortress that we humans have created.

5

Reliving the Past

I have been through some terrible things in my life, some of which
actually happened.
—Mark Twain

In *Hidden Lives*, a memoir of her family, the British novelist Margaret Forster describes herself in the first five years of her life in the third person. From the age of five, however, she switches to the first person. She explains the switch as follows:

> It was at this time, in 1943, when I was five, that my own real memory begins, real in the sense that I can not only recall actual events but can propel myself back into them, be there again in my Aunt Jean's room-and-kitchen, standing by the window at the back of the buildings, staring out at the outside staircase and the tops of the wash-houses, while behind me Jean asks what is the matter. . . . That is what I call real memory, not at all the same as "remembering" being taken to Ashley Street School to demonstrate my boasted ability to read. Though, because my mother later told me about it so often, I often claimed to remember it and could easily convince myself that I did.[1]

Up to that point in the book, she describes the life of a rather difficult little girl called Margaret as just another member of the family, but from then on everything is viewed through the lens of personal experience. The transition, as biography switches to autobiography, may even herald the arrival of the concept of self. It is the beginning of memory as a recursive phenomenon, with previous experience inserted into present consciousness.

My own conscious memory also begins at around five, when I trudged about a mile up a country road to my first school, accompanied by older boys from down the road whom my mother entrusted to look after me, but in fact they tormented me for wearing glasses. There are a few earlier episodes that I imagine I remember, but they may well be fabrications from things that were told to me. I often ask my undergraduate classes to declare their earliest memories. There are always a few souls who claim memories going back to infancy, even to birth, but most of the memories come up from age four or five. This pattern is corroborated in adult surveys of remembered events before the age of eight. Virtually no memories were recorded from the first three years of life, rising steadily to all adults remembering events from the first eight years.[2]

In the nomenclature of modern cognitive psychology, Margaret Forster's distinction between "real memory" and "remembering" is out of line. The term "remembering" is usually taken to refer to memory for actual events, located in time and space. This is also known as *episodic memory*. It involves consciously projecting oneself back in time, just as Margaret Forster did. The term "knowing" refers to the other kind of memory, also known as *semantic memory*, which is the storehouse of knowledge that we possess, but that does not involve any sense of conscious recollection.[3] I know that Paris is in France, but have no conscious memory of when I learned this. On the other hand I can vividly remember being in class as an eleven-year-old when I discovered that $(2n + 1)$, where n is any integer, was a general expression for an odd number. My memory is tinged with triumph, because we were asked to generate this expression ourselves, and I was the only one who did so.[4]

Episodic memory may combine with some aspects of semantic memory to make up what is known as *autobiographical memory*.[5] As the passage from Margaret Forster illustrates, it is often difficult to distinguish which parts of autobiographical memory are based on remembered episodes, and which on knowledge. In the study referred to above, the adults were not only asked to recall memories, but were also asked to describe things that they knew had happened to them, but could not actually remember. Knowing but not remembering early events depends largely on family folklore—stories, often repeatedly told, of childhood events, and often embellished through imagination to the point that one has the sense of re-experiencing them. Nevertheless "known" events declined steadily from birth to age eight, while "remembered" events increased from age two to age eight.

Endel Tulving has described remembering as *autonoetic*, or self-knowing, in that one has projected one's self into the past to re-experience some earlier episode.[6] Simply knowing something, like the boiling point of water, is *noetic*, and implies no shift of consciousness. Autonoetic awareness, then, is recursive, in that one can insert previous personal experience into present awareness. This is analogous to the embedding of phrases within phrases, or sentences within sentences. Deeper levels of embedding are also possible, as when I remember yesterday that I had remembered an event that occurred at some earlier time. Chunks of episodic awareness can thus be inserted into each other in recursive fashion. Having coffee at a conference recently, I was reminded of an earlier conference where I managed to spill coffee on a distinguished philosopher. This is memory of a memory of an event. I shall suggest later that this kind of embedding may have set the stage for the recursive structure of language itself.

Amnesias

Brain imaging suggests that there may be considerable overlap in the areas activated when people retrieve episodic and semantic memories, but each kind of retrieval also activates unique regions.[7] In cases of amnesia, which is the loss of memory caused by brain injury, it is episodic memory that usually carries the brunt of the loss. Semantic memory is relatively unimpaired. One well-known case is the musician Clive Wearing, whose plight has been aired in several television documentaries and in a book written by his wife Deborah.[8] Wearing was an acknowledged expert on early music and had built up a musical career with the BBC, when at the age of 46 he became ill with the herpes simplex virus. This effectively destroyed his hippocampus, a structure in the inferior temporal lobe that is critical to memory. Wearing's condition is well described by the title of a 2005 ITV documentary, *The Man with the 7-Second Memory*. That is, he has sufficient short-term memory to be able to respond to questions, and even converse, although he quickly forgets topics that he spoke about moments earlier.[9] He does remember some aspects of his life before the illness. For example he recognizes his children, but does not recall their names. Much of his semantic memory, though, remains intact. His vocabulary is largely unaffected, and he still knows how to play the piano or conduct a choir.

The most extensively studied case of amnesia for events is that of H. M., who underwent surgery for intractable epilepsy at the age of 27.[10] The surgery destroyed most of his hippocampus, along with surrounding regions, and left him in much the same state as Clive Wearing. He remembered some of his pre-illness life, but suffers memory loss reaching back in time from the operation, with better recall of earlier than of later events. This has led to the view that the hippocampus is responsible for laying down memories, and somehow holds them temporarily while consolidation takes place elsewhere in the brain. Loss of hippocampal function therefore not only prevents new memories from being formed, but also destroys past memories that are temporarily held there, before being distributed to other brain regions. The holding operation may last several years, with diminishing effectiveness, as the hippocampus relinquishes its grip.[11]

There has been some dispute about whether semantic memory and episodic memory are truly distinct. One might suppose, for example, that episodic memory is fragile simply because it is based on single events, whereas semantic memory is based on information that is repeated, and may be present in multiple episodes. Endel Tulving has undertaken careful studies of a particular patient, known as K. C., and shown that his difficulty with episodic memory relative to semantic memory cannot be attributed to the frequency with which the recorded information is presented. K. C. can recall factual information that he is unlikely to have rehearsed repeatedly, but cannot bring to mind events that lasted several days, such as being evacuated from home, along with tens of thousands of others, when a nearby derailment released toxic chemicals. This suggests that episodic memory can fail even when there is likely to have been repeated rehearsal. Tulving refers also to controlled experiments leading to the same conclusion.[12]

Tulving nevertheless argues that the storage of episodic memories depends on semantic memories that are already in place—one can scarcely record a visit to a restaurant without already knowing what a restaurant is, and what happens there—but are then related to the self in subjectively sensed time. This allows the actual experience of the event to be stored separately from the semantic system.[13] In this view, episodic memories could not be stored in the absence of semantic memory, which is perhaps why our childhood episodic memories do not begin until the semantic system is well established, by around age four or five. Nevertheless there is some evidence that episodic and semantic memories are more broadly dissociated. People with semantic dementia, a degenerative neurological disorder that afflicts some people in late adulthood, show severe decline in semantic memory, but their episodic memories remain remarkably and surprisingly intact.[14]

Unconscious Memory

There is another category of memory that appears to remain intact in cases of profound amnesia. This illustrated by the finding that H. M. could learn new skills, such as mirror drawing, where the task is to draw or trace a shape while viewing the hand and shape only in a mirror. This is a difficult task, because the tracing movements you see in the mirror are front-back reversed relative to the actual movements. H. M. improved with practice, just as people without memory deficits do, yet never remembered having done it before. He even showed consistent improvement on tasks that require mental operations, rather than learned movements. One example is the Tower of Hanoi task, in which rings are stacked in order of size on one of three pegs, with the largest ring at the bottom. The task is to move the entire stack to another peg one at a time, using all three pegs, but without ever placing a larger ring on a smaller one. Again, H. M. showed systematic improvement, although he never consciously recalled seeing the problem on previous attempts.

The unconscious memory that underlies these skills is called *implicit memory*, which enables us to learn without any awareness that we are doing so. It is presumably more primitive in an evolutionary sense than is *explicit memory*, which is made up of semantic and episodic memory. Explicit memory is sometimes called *declarative memory* because it is the kind of memory we can talk about or *declare*. Implicit memory does not depend on the hippocampus, so amnesia resulting from hippocampal damage does not entirely prevent adaptation to new environments or condition, but such adaptation does not enter consciousness. I am a poor typist, and could not consciously tell where the keys are placed on the keyboard, except perhaps for the rather touching [*sic*] fact that U and I are side by side, but my fingers find their way very quickly to the appropriate keys. This is implicit memory. I should add that I only use two fingers. The others know nothing.

Implicit memory is sometimes measured through a technique called *priming*. For example, if people are asked to recall a series of words they have been shown in the laboratory, their recall of individual words may be enhanced if they are preceded by some related word, or prime. For example, the word *animal* may prime recall of the word *aardvark*. Priming is remarkably resilient. In one study, for example, fragments of pictures were used to prime recognition of whole pictures of objects. When the same fragments were shown 17 years later to people who had taken part in the original experiment, they were able to write the name of the object associated with each fragment much more accurately than a control group who had not previously seen the fragments. Some of the people involved had no conscious recollection at all of the laboratory experiment 17 years earlier, and these people actually scored slightly higher on the priming measure than those who did remember the previous occasion.[15]

The Fragility of Memory

Implicit memory, then, is unconscious and seemingly invulnerable, whereas our episodic memories are fragile, even in the absence of brain injury. Indeed, we are not nearly as good at remembering events as we think we are. At one time it was thought that we store everything that we were ever conscious of, and that the reason that we seem to forget so much has to do with retrieval failure. That is, there are lots of things stored in memory that we can't find. It is true that retrieval is fickle—we all have the experience of being embarrassingly unable to remember someone's name when we need to introduce that person to a group of friends, only to remember it later when it no longer matters. More recent research suggests that memory failure is not just a matter of retrieval failure, which of course does occur, but also reflects failures of storage.[16] For this reason, a school reunion is a sobering experience.[17] At one such event, my school friends frequently recounted some apparently salient event that seemed to have completely escaped my own recollection, even though I was supposedly present when it happened. Symmetry was restored when I told of events that others seem to have forgotten. And of course some stories were embellished beyond recognition anyway. Some of my old mates, I do remember, were compulsive liars.

We now know that our memories are sieve-like; in fact, such is the length and complexity of our conscious lives that we retain probably only a tiny fraction. The émigré Czech writer Milan Kundera, in his novel *Ignorance*, put it like this:

> The fundamental given is the ratio between the amount of time in the lived life and the amount of time from that life that is stored in memory. No one has ever tried to calculate this ratio, and in fact there exists no technique for doing so; yet without much risk of error I could assume that the memory retains no more than a millionth, a hundred-millionth, in short an utterly infinitesimal bit of the lived life. That fact too is part of the essence of man. If someone could retain in his memory everything that he had experienced, if he could at any time call up any fragment of his past, he would be nothing like human beings: neither his loves nor his friendships nor his capacity to forgive or avenge would resemble ours.[18]

Well, perhaps that's an exaggeration—a hundred-millionth would amount to about 15 minutes' worth. Still, most of our conscious lives slip by without lasting trace.

The paucity of our memory for events may nevertheless be adaptive, since to remember everything that happens to us could be debilitating, clogging our minds with unwanted junk—like old clothes. Some individuals, in fact seem to suffer precisely because their memories are too good.

Can You Have Too Much Memory?

New Zealand children of my generation were brought up with a substance known as Marmite, a spread that is roughly the cultural equivalent of peanut butter in the United States. The jar contained the message "Too much spoils the flavor." To those not raised with this substance, the flavor is repugnant anyway, no matter how little you smear onto your slice of bread. The idea that too much might spoil the flavor must have been a poor marketing strategy, and the message no longer appears. Memory, perhaps, is a bit like marmite. Too much may spoil the mind.

There are indeed reasons to suppose that a good memory may be an impediment. A condition known as *savant syndrome* can result in prodigious powers of memory, but deficiencies in other aspects of intelligence. The most extraordinary case so far described is that of Kim Peek, who was the inspiration for the movie *Rain Man*.[19] Peek, known to his friends as "Kim-puter," began memorizing books at the age of 18 months and in his midfifties has now memorized 9,000 books. He has a vast storehouse of knowledge in history, sports, movies, space programs, literature, and Shakespeare, among other things. He has an extensive knowledge of classical music, and in middle life has even begun to play it. In common with other savants, he is an expert at calendar calculation, and is able to instantly tell you the day of the week for any given date. This is known to depend on massive memory.

Yet on a standard test of intelligence, Peek scored only 87 (the population average is 100). There was great variation, though, in the subscales that go to make up the intelligence score, with some in the superior range and others in the range of the mentally retarded. His brain is abnormal, being unusually large and lacking the corpus callosum, which is the main fiber tract that connects the left and right sides of the brain. Two other tracts that normally link the sides of the brain are also missing. He has an unusual sidelong gait, cannot button his clothes, and cannot handle the chores of daily life. He also has difficulty with abstract ideas. What this profile suggests is that a large and detailed memory can work to the disadvantage of other mental skills, and a memory that is too particular can impair ability to see relations and form abstractions. Too many trees, and it's hard to see the wood.

Another case in point was Solomon Shereshevskii, better known as "S," whose prodigious feats of memory were described by the Russian neuropsychologist Aleksandr Romanov Luria in his 1968 book *The Mind of a Mnemonist*. According to Luria, S's memory capacity was seemingly without limit, and he remembered trivial things for extremely long periods of time. For instance, he could accurately recall lists of words that Luria had presented 16 years earlier. His memory was largely visual, and when given verbal items he could transform them mentally, either by arranging them spatially, or using the "method of loci" whereby he would imagine them in familiar locations. The particularity of his memory was actually an impediment, because it prevented him from forming general concepts. He was unable to make sense of novels, since he would imagine scenes in precise detail, only to find his images contradicted at later stages. S was also a synaesthete—a condition in which inputs in one sensory modality give rise to sensations in another—so that spoken words were accompanied by visual sensations, such as "puffs" or "splashes," and a tone at 30 cycles per second and 100 decibels gave rise to "a strip 12–15 cm in width the color of old tarnished silver."[20]

Another case that has recently come to light is that of a woman known in the literature as A. J.[21] She would not be classed as a savant, but nonetheless has an unusual and intrusive autobiographical memory. She was born in 1965, but claims that if you give her any date between 1974 and the present she can tell you what day of the week it was, what she was doing on that day, and whether anything of great importance (such as the *Challenger* explosion of January 28, 1986) occurred. She kept a diary from age 10 to age 34, but her memory for events evidently does not depend on her rereading the diary, which she hardly ever does, nor does she consult a calendar before being asked questions about her life. She spends an excessive amount of time recalling her past, and tests have shown that what she does recall is extraordinarily accurate.

Paradoxically, though, when the researchers made a videotape of her and showed it to her a month later, she remembered very little of the video session itself, suggesting that her autobiographical memory is in fact very selective. Although she seems to remember something from every date, the actual amount remembered may be only a fraction of what actually happened. She claims to have had trouble memorizing historical dates, learning foreign languages or science, and got Ds in geometry. Her full scale intelligence quotient (IQ) is 93, a little below average, although she scored 122 on a standard memory test. Despite her academic shortcomings, she completed a bachelor's degree in social science, graduating at the age of 23.

Memory, whether exceptional or not, requires huge storage space. As mentioned earlier (in case you have forgotten), the average literate adult probably knows about 50,000 words, which implies that she knows about the same number of concepts, including objects, actions, qualities, professions, emotions—and on and on.[22] Episodic memories are generally made up of different combinations of otherwise familiar concepts, as in who did what to whom, when where and why. To accommodate all of this memory, it is not surprising that the human brain is large, even by primate standards. Something like 10 billion neurons exist in the human brain, each connected to other neurons—in fact, some neurons have as many as 30,000 connections. It's hard to get one's head around such numbers, even though they are in fact contained within the head. Given that memories are probably stored in terms of interconnections between neurons, one may be tempted to believe that the brain has infinite capacity. But the brain has lots of other things to do besides store memories. I suspect therefore that there is something of a trade-off between different mental pursuits, and obsession with some pursuits may prevent others from developing. Sometimes, as in the case of people with savant syndrome, the reason for the imbalance may be pathological—some defect of growth or hormone balance that allowed some faculties to develop at the expense of others. In other cases, perhaps, obsessions may arise in response to cataclysmic life events. And there is no doubt normal variation. Life would be dull if this were not so.

False Memories

> When I was younger I could remember anything, whether it had happened or not, but my faculties are decaying now, and soon I shall be so I cannot remember any but things that never happened.
> —Mark Twain

Although our memories of the past can be vivid, to the point that we seem to re-experience them, they can fail to be true to the original experience. Our memories of earlier exploits are often embellished or distorted, and often downright false—we "remember" events that simply never happened. The pioneer of research in this field, Elizabeth Loftus, recounts a false memory of her own mother's death:

> I remember a summer many years ago. I was fourteen years old. My mother, my aunt Pearl, and I were on vacation, visiting my uncle Joe in Pennsylvania. One bright sunny morning I woke up and my mother was dead, drowned in a swimming pool.[23]

The scene to her is vivid. In her mind she sees and smells cool pine trees, tastes iced tea, and sees her mother in her nightgown, floating face down. She cries out in terror, starts screaming, sees the police cars, lights flashing, and the stretcher carrying her mother's body. But the memory is false. She was in fact asleep when the body was discovered, not by her but by her Aunt Pearl. The memory is a construction, built partly of knowledge of what happened, and partly of extra details supplied by imagination.

Loftus studied false memory experimentally by giving people short narratives, constructed by family members, of events in their childhood. She then asked them for their memories of these events. One of the narratives described being lost in a mall, but was entirely false. About a quarter of her sample claimed to remember the event, and some even supplied details that had not been mentioned in the narrative. This study took place in 1991, and has been repeated many times with other false narratives that were nevertheless "remembered," such as being taken for a ride in a hot-air balloon, being attacked by a vicious animal, or being nearly drowned and rescued by a lifeguard. False memory is also studied using more prosaic methods, such as giving people lists of words to remember that omit one very obvious word; when asked to recall the list, participants will often insist that the missing word had actually been presented. For example, the list may include words like *bed, dream, rest, awake,* and so on, omitting the word *sleep*, but around half of the participants later remember the word *sleep* as having been in the list.[24] This experiment has become something of a cottage industry in experimental psychology, and is a regular feature of undergraduate laboratories.

Actually, false memories were widely discussed in the late nineteenth century, and were known as paramnesias.[25] An especially horrific example is described by the hypnotherapist Hippolyte Bernheim, who recounts how he suggested to a patient that she had witnessed through a keyhole an old man raping a little girl, who struggled, was bleeding, and was then gagged. His suggestion concludes "When you wake up you will think no more about it. I have not told the story to you; it is not a dream; it is not a vision I have given you during your hypnotic sleep; it is truth itself." Three days later Bernheim asked a friend and distinguished lawyer to question the patient as though he were an examining judge. The patient recounted the events in detail as suggested to her, and even when encouraged to doubt them maintained the truth of the testimony "with immovable conviction."[26] Today, fortunately, ethical regulations would forbid such a study.

In recent times, the question of false memory has been revived and has assumed special importance. In the 1980s and 1990s, many therapists adopted the view that psychological problems in adulthood could be traced to sexual abuse during childhood. Because of their traumatic nature, such memories were often repressed, and the main purpose of therapy was to recover these memories, so that patients could then face the real causes of their problems and deal with them—presumably with the therapist's help. The most extreme expression of this view was a book by Ellen Bass and Laura Davis entitled *The Courage to Heal*, which was first published in 1988 but has since gone through several editions. Bass and Davis, who had no formal training in psychology or psychiatry, were nevertheless bold enough to tell their readers:

> If you don't remember your abuse, you are not alone. Many women don't have memories, and some never get memories. This doesn't mean they weren't abused.

In another notorious quote from the book, Bass and Davis write, "If you think you were abused, and your life shows the symptoms, then you were." This statement commits the logical fallacy of affirming the consequent. It is undoubtedly true that childhood abuse can cause later symptoms of psychological distress, but this does not mean that psychological stress must have resulted from childhood abuse. Murder results in death, but this does not mean that death is always due to murder. Unfortunately, widespread acceptance of Bass and Davis's edict too often led to aggressive therapy designed to help distressed people recover the memories of the abuse that led to their distress.[27] This led to the recovery of "memory" for events that did not happen, just as Loftus's experimental subjects were never in fact lost in a mall. The whole issue of recovered memory versus false memory became intensely political in the 1990s, with many cases of men being given jail sentences for abuse despite being entirely innocent;[28] one such case in my own country is described in detail in a prize-winning book.[29]

An Evolutionary Perspective

By now it will be apparent that memory is not quite what it seems. It is far from a faithful record of past events, and is also complex, made up of several systems. This is an appropriate point, then, to consider the evolution of memory systems, and how they might contribute to the adaptive fitness of organisms. I shall suggest that there is a hierarchy of memory systems, each contributing in a special way. I will here take "memory" to mean any adaptation involving some change in an animal's behavior that results in increased fitness.

A good proportion of behavioral adaptations are instincts, which are "memories" only in the sense that they refer to adaptive changes that took place in earlier generations, and were transmitted genetically. Genetic endowment is not sufficient, though, and instinctive behaviors will often not develop unless the environmental conditions are adequate. Instincts can be quite complex, and include activities such as seasonal migrations of birds and animals, hoarding and caching, dam-building in beavers, and so forth. These are effective adaptations to predictable seasonal events, and are effective so long as seasonal changes are indeed regular. Such instincts can be disrupted by climate change, or by the predations of other species, especially humans, with our unparalleled capacity to wreak destruction, not only on the geography of the planet, but also on seasonal change itself.

Of course we humans also have lots of instincts, such as emotions, sexual drive, and, one may think, a predisposition to gossip. Steven Pinker has maintained that language itself is an instinct;[30] given normal development all humans talk or sign, some of them too much. Again though it does depend on the environment, and a child brought up without human contact will not develop language, as illustrated by the famous case of Genie, a girl who was isolated until the age of 13 and never learned proper language.[31] Of course the particular languages we learn also depend on the particular linguistic environment we are brought up in; I fear I will never learn Chinese. The development of birdsong involves similar principles. Although fundamentally instinctive, the song of the chaffinch requires that the young bird be reared where it can hear the song of other chaffinches.

Learning offers a more flexible means of adaptation, since it allows adjustment to environmental events or contingencies that occur in the individual's own lifetime. One form of learning is classical conditioning, made famous by Ivan P. Pavlov's demonstration that if a bell is sounded prior to the presentation of food, animals soon salivate to the bell.[32] More generally, classical conditioning may serve to generate appropriate emotional or anticipatory responses to situations, animals, or objects that pose threats or promise rewards. Classical conditioning depends on the involuntary elicitation of bodily events, such as salivation, blushing, or emotional responses like fear and anger, which then become attached to previously neutral events.

Pavlov's experiments were carried out with dogs. The most famous (not to say notorious) study of classical conditioning in humans was carried out by the founder of behaviorism, John B. Watson. The subject[33] was a nine-month-old orphan boy known as Little Albert.[34] Watson first determined that Little Albert was not afraid of objects, such as a rat, rabbit, monkey, dog, or mask, but that he was afraid of a loud noise, created by banging a metal bar with a hammer. By presenting the rat and the noise at the same time, Watson was able to condition Little Albert to be afraid of the rat alone—and incidentally afraid of all the other objects as well. Little Albert was adopted shortly after these unfortunate experiences, and Watson did not have the opportunity to recondition him not to fear these objects. One must feel sympathy for both little Albert and his adoptive parents, and as in the earlier study by Bernheim, no ethics committee nowadays would permit such an experiment to be repeated.

Another form of conditioning, known as operant conditioning, involves the shaping of behaviors that are emitted rather than elicited, and might therefore be considered voluntary rather than involuntary. Such behaviors might include a rat pressing a bar, a pigeon pecking a disk, or a human operating a gambling machine in a casino. In operant conditioning, such behaviors are progressively controlled, or "shaped," by the delivery of rewards and punishments. They can also be brought under what operant psychologists call stimulus control; if otherwise neutral stimuli or events are associated with the rewards and punishments, they too can shape behavior. The dinner bell, then, may not only induce salivation, but may also serve as the signal to enter the dining hall. The behaviorist B. F. Skinner sought to explain all human behavior, including language, through the principles of operant conditioning,[35] leading to the somewhat robotic view of human society and behavior described in his utopian novel, *Walden Two*.[36]

These different forms of conditioning may be said to fall into the category of implicit memory. As we have seen, implicit memory may also include the learning of skills and even mental strategies to cope with environmental challenges. Implicit memories are elicited by the immediate environment, and do not involve consciousness or volition. Of course one may remember the *experience* of learning to ride a bicycle, but that is distinct from the learning itself. Many of my own childhood memories involve bicycles, including being launched by one parent in the direction of the other, turned around and launched again, thus riding in wobbly fashion back and forth. I still bear the scar from a later fall, which I remember vividly. I was nine years old, and I can remember the doctor's amusement, as he stitched up the gash in my knee, when I asked him if the wound would be fatal.[37] These are episodic memories, independent of the process of actually learning (more or less) to ride the bike.

Compared to implicit memory, explicit memory provides yet more adaptive flexibility, because it does not depend on immediate evocation from the environment. That is, we can consciously bring to mind such memories at any time, as when we muse about what happened yesterday, or try to bring to mind specific places or people that we know. This is not to say that the environment is unimportant, since we are constantly reminded of events by happenings in the world around us, or by probing questions, such as *What on earth were you up to last night?*

As we have seen, explicit memory includes both semantic and episodic memory. Our semantic memories are vast storehouses of facts about the world, including physical knowledge of the world and the particular environments that we inhabit, knowledge of our friends and how they behave, and even some knowledge of how language works, including vocabulary, some rules of grammar, and pragmatics—the different ways we can use language to achieve social goals. Language, though, is complex, and grammar for the most part depends on implicit rules—even linguists haven't figured them all out explicitly. By and large, semantic memory is the stuff of education, and is continually being upgraded and expanded through the processes of science and technological invention. Sometimes, it all seems too much.

Episodic memory, then, sits at the top of an evolutionary pyramid, providing the ultimate in fine-tuning our personal lives. By recording specific events, we can react more precisely to similar events in the future, and plan future activities in detail. We can also intentionally sift through different episodes in our lives to extract relevant information—perhaps in a fashion analogous to the insertion of phrases into sentences, or sentences into narratives. By knowing what went wrong on previous occasions, we can avoid a similar catastrophe in the future. On this view, then, episodic memory is not so much a permanent record of past events as a series of YouTube–like sequences, sometimes embellished or distorted, to provide a stock of situations that we can use to construct viable and detailed futures. Clearly it would be wasteful of storage space to record every event that we experience, since our daily lives are full of repetition. Repetitive events may well contribute to semantic memory, the generalities of the world, but episodic memory is concerned with salient particularities. That said, it is still not clear from research precisely which episodes are likely to be remembered and which are not. Episodic memory has a fickleness about it.

But there is more to the story than the examination of the past. As we shall see in the next chapter, episodic memory may also provide us with a sense of time, extending into the future as well as the past.

6

About Time

"It's a poor sort of memory that only works backwards,"
the Queen remarked.
   —Lewis Carroll, *Through the Looking Glass*

One important aspect of episodic memory is that it locates events in time. Although we are often not clear precisely when remembered events happened, we usually have at least a rough idea, and this is sufficient to give rise to the general understanding of time itself. Episodic memory allows us to travel back in time, and consciously relive previous experiences. Thomas Suddendorf called this *mental time travel*, and made the important suggestion that mental time travel allows us to imagine future events as well as remember past ones.[1] It also adds to the recursive possibilities; I might remember, for example, that yesterday I had plans to go to the beach tomorrow. The true significance of episodic memory, then, is that it provides a vocabulary from which to construct future events, and so fine-tune our lives. Natural selection cannot operate on the basis of past recollection per se, but only on what it contributes to survival in the present and future.

There is growing support for this suggestion. What has been termed *episodic future thinking*,[2] or the ability to imagine future events, emerges in children at around the same time as episodic memory itself, between the ages of three and four.[3] Patients with amnesia are as unable to answer simple questions about past events as they are to say what might happen in the future.[4] Clive Wearing, the amnesic musician described in the previous chapter, constantly complains that he has no past, and is stuck in the present. He is forever under the impression that he has just woken up. His plight is

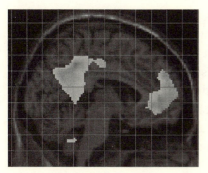 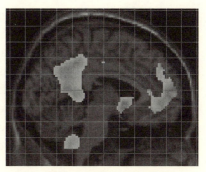

Past event > control   Future event > control

Figure 9. Midsagittal fMRI scans of brain activation elicited by imagining past events (*left*) and future events (*right*). Activated areas are shown as white. (Reprinted from Addis et al. 2007, with permission from Elsevier.)

captured in the title of his wife's book, *Forever Today*. Amnesia for specific events, then, is at least in part a loss of the awareness of time.

Brain-imaging studies confirm the close connection between episodic memory and imagining the future. Indeed the idea of mental time travel was anticipated in some of the earliest studies of cerebral blood flow by the Swedish physiologist David Ingvar, who concluded that activation in the prefrontal cortex was especially important in providing the internal connection between past and future. He wrote:

> On the basis of previous experiences, represented in memories, the brain—one's mind—is automatically busy with extrapolation of future events and, as it appears, constructing alternative hypothetical behavior patterns in order to be ready for what may happen.[5]

Using more refined techniques based on functional magnetic resonance imaging (fMRI), Daniel L. Schacter and his colleagues at Harvard University have shown that parts of the medial temporal lobe, as well as the prefrontal cortex, are activated when people are prompted to remember past events or to imagine future ones. Schacter and colleagues refer to this as a core network.[6] But of course people can usually distinguish remembered past events from imagined future ones, and some studies have shown in several brain areas greater activation in response to future than to past events,

perhaps indicating that imagining the future requires more inten-
sive imaginative construction.[7] Some studies, though, have shown
greater activation in recalling past events than in imagining future
ones,[8] so the evidence is inconsistent. Reliving the past also involves active construction,[9] sometimes to the detriment of veracity,
and I suspect that people may occasionally even confuse imagined
future events with those that actually happened. The brain itself
seems to hardly know the difference.

Is Mental Time Travel Unique to Humans?

He said "What's time? Now is for dogs and apes! Man has
Forever!"
—from *A Grammarian's Funeral* by Robert Browning

Like Browning's grammarian, Suddendorf and I also claimed that
mental time travel, including episodic memory, is uniquely human.[10] The same idea was also suggested by the German psychologist Wolfgang Köhler. Köhler was working at a primate research
facility maintained by the Prussian Academy of Sciences in the Canary Islands when World War I broke out. He was marooned, and
occupied his time studying the behavior of nine chimpanzees contained in a large outdoor pen. His work is famous for showing that
chimpanzees sometimes solve mechanical problems through the
use of insight rather than mere trial and error.[11] Köhler nevertheless concluded that, for all their improvisational skills, chimpanzees had little conception of past or future. Chimpanzees are our
closest nonhuman relatives, so Köhler's observation suggests that
mental time travel evolved at some point after the split between
the hominin and ape lineages.

The idea that mental time travel is uniquely human has nevertheless been challenged. The most serious challenge comes, not
from apes, but from birds. Some show prodigious memory for the
locations in which they have hidden food for later consumption.
Clark's nutcrackers seem to be among the most prolific, storing
seeds in thousands of locations and recovering them with high (but
not perfect) accuracy.[12] This need not mean, though, that they remember actually caching the seeds—they may simply know where

the seeds are. It has been suggested that the case for episodic memory would be strengthened if the birds could be shown to have some memory for *when* the seeds were cached, implying memory for time as well as place. Memory that includes storage of *what*, *where* and *when*, dubbed *www memory*,[13] has been taken by some researchers as sufficient evidence for episodic memory, and therefore of mental time travel, at least into the past.

The challenge to demonstrate www memory in birds has been taken up by Nicola Clayton and her colleagues at Cambridge University, using scrub jays. In clever experiments, Clayton allowed the birds to cache food in different compartments of ice-cream trays.[14] They were given two different foods to cache, worms and peanuts. They prefer worms to peanuts, and went to the location where worms had been cached rather than to where the peanuts were hidden, but only if the worms had been recently cached. If too much time had elapsed, and the worms were likely to have decayed and become inedible, they went instead to where the peanuts had been cached. This suggests that they remembered not only *where* they had hidden a specific food, but *when* they had hidden it. This implies further that they might have been traveling mentally back in time to the point at which they had originally cached the food.

These birds are also apt to steal each other's cached food. If a bird that had itself stolen food was watched while caching its own food, it later privately recached it, presumably in order to prevent the watcher from stealing it. It takes a thief to know a thief. Recaching also depended on which bird was watching. Clayton and her colleagues have found that recaching was more likely to occur if the predatory watcher was a dominant bird than if it was a subordinate bird.[15] Thus we might add a further *w*—the birds seem to know what, where, when, and *who*. Their behavior also seems to suggest mental time travel into the future as well as into the past, since recaching might be taken to imply anticipation of a future theft.

Meadow voles, it has been claimed, can also remember what, where, and when.[16] In an experimental study, male voles were first allowed to explore two chambers, one containing a pregnant female 24 hours prepartum, and the other containing a female that was neither lactating nor pregnant. Twenty-four hours later, they were again given access to the chambers, which were now empty and clean, and spent more time exploring the chamber that had contained the previously pregnant female than the one that had housed the other female. This suggests that they had remembered the pregnant female, and understood that she would now be in postpartum estrus, a state of heightened receptivity. In another condition they first explored a chamber containing a female in postpartum estrus and another containing a female that was neither lactating nor pregnant, and was not in estrus. Twenty-four hours later they were again allowed to explore the cages, now clean and empty, and showed no preference for the chamber that had housed the female in estrus. This suggests that they realized the female would no longer be in a state of heightened receptivity.

Nevertheless it is perhaps too early to conclude that these animals—birds or voles—can really travel mentally in time in the way that we humans do. The retrieval of worms or nuts depending on how long they have been cached need not imply that the jays actually remember the act of caching, as distinct from knowing when and where the food was cached. I know that Captain James Cook reached New Zealand in 1769, but I do not remember it; indeed I wasn't even born at the time, despite what my students may think. I don't even remember being told it. In the case of the birds, a simple time tag attached to the memory, as a kind of "use-by date," could be sufficient. Similarly the voles' understanding of when a female is likely to be receptive could be based on knowledge of ovulation cycles, with a variable time marker attached to the memory. A distinction can be drawn between knowing *when* something happened and knowing *how long ago* it happened, and a recent study shows that rats, in choosing where to find food in a maze, are governed by the latter, not the former.[17] Rats don't travel mentally in time, and there is still no strong reason to suppose that scrub jays or voles do either.

And do the scrub jays recache their food when a likely thief is watching because they actually envisage a future act of theft? Not necessarily. Their behavior might simply be learned through the association of a watcher and subsequent loss of food.

But we are in difficult territory here, since psychological concepts are slippery, and it is often all too easy to explain a set of results in different ways to suit one's theoretical predilections. Even if birds or voles can be said to have a degree of mental time travel, the context in which it operates is narrow. Scrub jays are specialists in the caching and recovery of food, and their behavior in this context has evolved subtleties that may not extend to more general activities. And all animals are specialists in mating. Somehow I doubt that they can lie back and reminisce about the past or dream about the future, as we humans do.

One theory that affords a potential test of mental time travel into the future is the so-called Bischof-Köhler hypothesis, which states that only humans can flexibly anticipate their own future mental states of need and act in the present to satisfy them.[18] We go to the supermarket to buy food even though not hungry at the time, or attend expensive and often dull university courses in the interests of a later lucrative career. There have been several attempts to refute the hypothesis that such behaviors are unique to our species. In one study, for example, scrub jays, given the choice of two foods to cache, ceased to cache one of those foods if it would be available later when they were hungry.[19] However, this behavior could have been based on learning rather than on the anticipation of which food would be available later, when hunger struck[20]—there seems no good reason to suppose that the birds traveled mentally in time to a future occasion when they would be hungry. In all cases so far described, processes simpler than mental time travel can account for the findings.

The manufacture of tools for specific purposes is also sometimes taken as evidence for mental time travel into the future. Here, I'm afraid, we must yet again defer to birds, since New Caledonian crows are able to manufacture tools from twigs and bits of wire to solve mechanical problems, although chimpanzees make tools of at least comparable complexity.[21] Of course many examples of tool use may be a simple matter of improvisation to solve immediate problems, rather than planning for a more distant future. More convincing evidence of future thinking comes from the crows' ability to shape leaves from pandanus trees. These leaves have spikes along one side, and are therefore useful for inserting into holes containing grubs, which are then caught on the spikes and extracted. The crows shape these tools in tapered fashion, so that the wider end is held in the beak and the narrow end inserted into the hole. Given the ubiquity of these tools, the standardization of design, and the evidence for cultural transmission of technique,[22] it is highly unlikely that they are simply the outcome of improvisation. These tools are planned.

Chimpanzees also show many cultural variations, in the way in which they use tools as well as in other behaviors.[23] And some groups of chimpanzees do store hammers and anvils for years of use in cracking nuts, although this need not mean that they actually imagine themselves using these tools in the future.[24] Nevertheless there has been a recent claim that orangutans and bonobos, at least, save tools not needed in the present for use up to 14 hours later, which might suggest mental time travel,[25] although it is not entirely clear that the animals were not responding simply on the basis of past associations, rather than actively imagining a future event.[26] The bonobo Kanzi is said to be able to lead someone to a location where he knows something to be located, but again this need not imply that he remembers the act of visiting that location previously. The critical distinction between knowing and remembering is actually very difficult to test in nonhuman species.

In humans, at least, mental time travel implies the conscious acting out of episodes, whether past or future, which further suggests recursion. That is, a conscious episode is embedded in present consciousness. This can proceed to deeper levels, as when I remember that yesterday I planned an episode—perhaps a dinner party, for some date in the future. It may be this recursively embedded structure that differentiates our own time-governed behavior from that of other species.

Time and the Human Condition

We might concede that there is limited ability for remembering past events in other species, and perhaps even for imagining future ones, but this pales beside the extraordinary impact that mental time travel has had on the human condition. Manufactured stone tools go back some two million years in hominin evolution, and

Figure 10. Descartes discovers he exists through time (author's drawing).

probably reflect an emerging understanding of the concept of time. Events located in time prey heavily on our conscious lives, whether in reminiscing, regretting past mistakes and hoping to eradicate them in the future, planning dinner parties and weddings, imagining what life will be like after retirement. We are ruled by clocks, calendars, diaries, appointments, anniversaries—and taxes. Indeed our concept of time now extends back to the very origin of the universe, which came about, we are told, in a Big Bang somewhere around 13.7 billion years ago.[27] This claim of course leads to metaphysical anxiety about what was going on before that.

All species are driven to some extent by time, as in seasonal changes or the daily cycle of the sun, but these depend on bodily rhythms as well as on external signals, such as temperature or ambient light. Humans have measured time in deliberate fashion, to guide intentional behaviors. We measure it in seconds, minutes, hours, days of the week, months, years, centuries, millennia, epochs, eras, eons. These can be understood recursively, in that each element of time is embedded in a larger one—like the smaller fleas on the backs of larger ones in Augustus de Morgan's poem. Time can also be understood as cyclic rather than linear.[28] Seconds within minutes,

minutes within hours, hours within days, days within weeks, weeks within months, months within years, and so forth, all circle around in a dazzling array of wheels within wheels. Explicit labels, such as clock times, specific days of the week, or months of the year, can help us locate specific events in time, independently of bodily rhythms or seasonal fluctuations.

As I suggested in chapter 1, it is perhaps through iteration rather than true recursion that we understand the nature of time. Just as tomorrows extend indefinitely into the future, so yesterdays extend indefinitely back into the past. This generalized notion of time comes with costs, such as a lingering sense of guilt for remembered indiscretions, or anxiety over anticipated events that might repeat earlier unpleasant ones. The most fateful consequence of mental time travel, though, may be the understanding that we will all die. The lines of David Watts's hymn "Oh God our Help in Ages Past," based on Psalm 90, tell the truth we all must face:

> Time like an ever rolling stream,
> Bears all its sons away,
> They fly forgotten as a dream
> Dies at the opening day.

The understanding of death, along with theory of mind, is of course an unpalatable aspect of the human condition, but it is moderated to some extent by the understanding that time continues beyond death. Our lives, perhaps, can be continued in those of our offspring. One function of religions, though, might be to instill the notion that life itself can continue after death, whether in the form of Heaven, Hell, Valhalla, or Nirvana. This belief in life everlasting may offer comfort through the promise of rewards after death, thereby providing some consolation from the sheer bleakness that death otherwise suggests. It can also be used to manipulate behavior, as in threats of hellfire and damnation if we do not conform to certain rules of behavior, or heavenly rewards if we do. Acts of terrorism through suicide, as in the destruction of the Twin Towers in New York on 11 September 2001, or Japanese kamikaze pilots who died for their emperor in World War II, might be considered extreme examples of such manipulation, although one might argue that all soldiers who submit to warfare are similarly manipulated, and celebrations of the glorious dead continue long after a war is over. The offer of life after death, with associated rewards and punishments, is remarkably ingenious, since there seems no way in which we can be either gratified or disappointed—at least if these emotions are restricted to the living. This may explain the origin of faith.

The relation of religion to the sense of time is born out by the Pirahã, the small Amazonian community that we encountered in chapter 2. It was the Pirahã's lack of a sense of time, and consequent failure to understand what religion is all about, that led Daniel Everett to forego his career as a missionary among them and become an atheist—and a professor.[29]

The Pirahã notwithstanding, we can trace something of the prehistory of mental time travel through burial sites. There is some evidence that the Neandertals buried their dead, although the reasons for burial may have been more practical than religious. In some early human burials, though, symbolic material is added to the grave, suggesting that those who buried their dead had some notion of a spiritual life that continued after death of the body. At least from the evidence to date, such burials were restricted to our own species, *Homo sapiens*.[30] Perhaps the earliest example is from a burial site at Qafzeh Cave in Israel, in which the head of a deer was placed on the body of a child who was buried. This burial is dated at about 100,000 years ago, and is associated with early anatomically modern humans.[31]

The understanding of time is of course not all bad news. Sometimes the knowledge that time will pass is a consolation, as when Viola in Shakespeare's *Twelfth Night*, masquerading as a man, finds herself in an impossible situation, and is moved to say

> O Time, thou must untangle this, not I;
> It is too hard a knot for me t'untie!

And it is arguably the understanding of time that has shaped human destiny. Instinctive behaviors like dam-building in beavers or migrations in birds are future-oriented, but in limited and inflexible ways, at least compared to the extraordinary way in which we humans can plan our futures, or manufacture buildings, transportation devices, household conveniences, suitable clothing, and impressive CVs, that help us guarantee future prospects.

According to Ayurveda, the ancient Indian science of life, "You are what you eat." Much more obviously, though, we are what we remember, and what we plan. Mental time travel, then, provides us with the concept of a conscious self that extends in time. The psychologists Hazel Markus and Paula Nurius wrote of "possible selves," derived from representations of the self in the past and including representations of the self in the future.[32] The pioneering psychologist William James expressed much the same idea when he wrote of "potential social Me" as distinct from "immediate present Me" and "Me of the past."[33] The concept of different possible selves provides the primary motivation that guides our future endeavors. As Markus and Nurius put it, "I am *now* a psychologist, but I *could* be a restaurant owner, a marathon runner, a journalist, or the parent of a handicapped child."[34] The motivation can operate both positively and negatively; I can imagine myself as a roaring success, whether at parties, on the rugby field, or in scientific achievement, or I can see myself as a dismal failure.

Mental Time Travel and Fiction

The concept of mental time travel helps explain the fragility of episodic memory, documented in the previous chapter. The importance of episodic memory, then, lies not in providing a detailed record of the past, which it does very poorly, but rather in its role in constructing future scenarios. As we have seen, episodic memory itself is essentially a construction; Ulric Neisser recently wrote, "Remembering is not like playing back a tape or looking at a picture; it is more like telling a story."[35] It is a process whereby we establish our own identities, often in defiance of the facts.

This leads to fiction itself. The same constructive process that allows us to reconstruct the past and construct possible futures also allows us to invent stories. We humans are addicted to folktales, legends, novels, movies, plays, soap operas, and everyday gossip. It is the power of recursion that makes these things possible. Critical to all of them is language, the device that enables us to share our memories, future plans, and dreams. More of that, though, in the next chapter.

7

The Grammar of Time

Jane Goodall, who knows chimpanzees better than anyone else does, was recently asked how close chimpanzees are to humans. She replied:

> What's the one obvious thing we humans do that they don't do? Chimps can learn sign language, but in the wild, so far as we know, they are unable to communicate about things that aren't present. They can't teach what happened 100 years ago, except by showing fear in certain places. They certainly can't plan for five years ahead. If they could, they could communicate with each other about what compels them to indulge in their dramatic displays. To me, it is a sense of wonder and awe that we share with them. When we had those feelings, and evolved the ability to talk about them, we were able to create the early religions.[1]

So far, there is little convincing evidence that animals other than humans are capable of mental time travel—or if they are, their mental excursions into past or future have little of the extraordinary flexibility and broad provenance that we see in our own imaginative journeys. The limited evidence from nonhuman animals typically comes from behaviors that are fundamentally instinctive, such as food caching or mating, whereas in humans mental time travel seems to cover all aspects of our complex lives.

One problem, though, is that it is relatively easy to assess mental time travel in humans through the use of language, but there is so far no clear means of assessing it in nonhuman animals. As we saw in the previous chapter, one suggested criterion for the assessment of episodic memory, or mental travel into the past, is the www criterion; that is, an animal might be said to have demonstrated episodic memory if its behavior showed recollection of *what*, *where*, and *when*. I argued that this does not quite do the trick, as we can know the whats, wheres, and whens of events without necessarily reliving them in our minds. In terms of the sheer particularity of our remembered episodes, Steven Pinker was probably closer to the mark with a further flowering of *w*s: "who did what to whom, what is true of what, where, when and why."[2] More intangible, though, is the subjective quality of episodic memory, the sense that we are mentally reliving the past. There is as yet no established way of measuring that in another species. The problem is compounded if we add mental time travel into the future. Our mental lives are stretched in time, whereas other animals, as far as we know, live mainly in the present.

The difficulty of showing that nonhuman animals do have episodic memories or imagined futures, then, stands in marked contrast to the ease with which we can probe these capacities in humans. Indeed it is often difficult to prevent people from divulging their past exploits, whether a recent trip to Spain or a hole-by-hole account of a round of golf. Similarly, our friends are often only too ready to divulge their plans for some future business venture or romantic conquest. In the evolution of our species, then, language and mental time travel seem to be linked. Language may have evolved precisely so that we can share our mental travels through time, and the absence of language in other species may stem from the absence of mental time travel itself.

Despite the tedium of hearing in detail about someone else's bridge evening, the sharing of our pasts and futures is almost certainly adaptive. Even golfers may benefit from golfing stories—it's only the duffer who must occasionally suffer. If episodic memory evolved to enable us to plan our futures, and choose between alternatives, then we can only benefit if we add the experiences of others to our own. We humans seem to have an inordinate interest in the lives of others, whether from the gossip of our friends or the tabloid press. To be valuable, moreover, episodes need not be based on fact, and most of us are also addicted to television soaps, operas, movies, novels—even just-so stories.[3] Skilled storytellers can create fictional episodes that tell us much about the human condition. The universal predilection for murder mysteries, for example, may help us understand the vagaries, and perhaps deviousness, of human behavior—but more of that later. As a graduate student in psychology at McGill University in Montreal, I took a compulsory seminar with the legendary Donald O. Hebb, Canada's most famous psychologist, who used to tell us that you could learn more about human relations from reading novels than from taking courses in psychology.[4] (I stayed with the course.)

I think that grammatical language evolved primarily to enable us to share episodes, thus greatly enlarging the vocabulary of real-world happenings for the construction of personal futures.[5] This provides a vast resource of information about how people behave. Since this information is shared, it also creates cultural expectations and templates for future planning. Of course language is also used to communicate semantic information, as indeed I am attempting to do in this book, but the initial impetus, I suggest, came from the emergence of episodic memory, mental time travel, and the sense of time. And although episodic memories may require a foundation of semantic memory,[6] semantic memory in turn is no doubt shaped and extended through the accumulation of episodic memories.

Several of the critical properties of language, then, probably evolved from the relaying of events, whether past, present, future, or fictional, most of them located at times other than the present, including imaginary time. Other times also means other places, since we are peripatetic creatures, restlessly moving about the planet— and occasionally off it. Language is exquisitely designed to communicate "who did what to whom, what is true of what, where, when and why," to quote Pinker again—this time in the context he intended.

What, then, are the properties of language that allow us to do this?

Symbolic Representation

The first requirement is that we need ways to refer to objects, actions, qualities, and so forth that are not physically present, and cannot be referred to by pointing, or by shared attention—as when two people are both looking at the same painting, or listening to the same piece of music. That is, referring to the not-physically-present requires symbols. This in turn requires double coding. We need first a set of internally stored concepts, and second a list of communicable symbols to refer to them. It does not really matter what the symbols are, so long as they are sufficiently distinct to avoid confusions among them.

In chapter 4, I referred to Ferdinand de Saussure's view that language is by definition composed of signs that are arbitrary, rather than iconic or onomatopoeic. I argued there, contrary to Saussure, that the use of arbitrary symbols derives from practical concerns rather than linguistic necessity. Relative to speech, signed languages allow greater use of symbols whose form gives a physical clue to what they refer to, since signed languages operate in a visual medium, and most of the things we talk about, whether objects or actions, can be described visually rather than acoustically. But even in signed languages the symbols tend to become conventionalized and lose their pictorial aspect. And although a picture may be worth a thousand words, pictures take too much time to construct for efficient communication, especially with our limited bodily resources. The associations we form between symbols and concepts are also sufficiently rich to allow the symbols themselves to be a kind of shorthand. In referring to different kinds of dog, for example, labels are sufficient to conjure up the appropriate images; the mind, not the symbol, does the work. Thus the words *beagle, dachshund,* and *collie* bear no relation to these varieties of dog, or even to each other, but you can picture what each of them looks like, and if you're a dog lover conjure up their other characteristics.

In the interests of speed and efficiency, moreover, more frequent words tend to be shorter. This is captured by Zipf's law, which states that the length of a word is inversely proportional to its rank in frequency. The reason for this is evident from the title of Zipf's 1949 book, *Human Behavior and the Principle of Least-Effort.* Language calibrates itself so as to require as little effort as possible, and words tend to grow progressively shorter as they become more common. Hence we have the progression from *television* to *telly* to *TV,* or in my own country from *university* to *varsity* to *uni.*

Of course with modern technology, pictures can indeed replace words or signs, at least in some contexts. A video of one's vacation in Europe may even be more effective than a verbal account, but does not provide the to-and-fro possibilities for conversation. A more serious threat, perhaps, comes from texting, with its increasing short-cuts and rapid ways of taking and sending digital photos. Texting restores the advantage of being able to communicate at a distance to the hands, at the expense of the mouth, and may well soon become the conversational medium of choice. Thus if language progressed from hand to mouth, as I argued in chapter 4, we may now be witnessing a switch back again.

As we saw in chapter 3, the use of symbols is not exclusive to humans. Great apes have been taught to communicate using gestures or visual symbols (constructed by humans), and chimpanzees in the wild emit pant hoot calls to signal the discovery of food. Vervet monkeys use different calls to warn of different predators. But we humans are unmatched in the number of symbols we use. The time dimension vastly increases the mental canvas, since reference to different times generally involves different places, different actions, different actors, and so on. Steven Pinker estimates that the average literate person stores about 50,000 concepts,[7] and requires the same number of words in order to refer to them. There are probably qualitative differences as well. Terrence Deacon refers to us humans as "the symbolic species,"[8] and points out that the symbols we use have been wrested away from simple association with their referents, and are used flexibly in human thought. Thus words can bring to mind other words, and can be used freely in the absence of what they refer to. Again, this is due in part to the human sense of time, and the ability to use symbols to evoke events from the past, or imagined future events.

A clue to the origin of reference to absent entities comes from a study comparing 12-month-old infants with chimpanzees.[9] In different but comparable settings, the babies and the chimps first repeatedly watched a human adult place desired objects (toys for the babies, food for the chimps) on one platform, and undesired objects (paper towel for the babies, bedding material for the chimps) on another similar platform. The participants were then tested under two conditions. In one, the adult brought out the desired object and placed it under its platform so the participant couldn't actually see it. The majority of both the babies and the chimps pointed to the platform where the object was hidden, thereby prompting the adult to give them the object. In the second condition, the platforms were left empty. The majority of the babies still pointed to the platform where the desirable object had been, but the chimpanzees did not, despite evidently wanting to get food. Even by one year of age, human infants may have beginnings of displaced reference; or the ability to refer to objects that are not present, but chimpanzees, our closest nonhuman relatives, apparently do not.

Marking Time

Languages have also evolved means of indicating the time at which an event took place. In many languages this is achieved by marking verbs according to tense. This can be more complex than merely indicating past, present, or future. We can also distinguish perfect tenses, referring to actions that are completed relative to the present, from the so-called imperfect ones, which refer to actions not yet complete. Thus *I had been to town* is the past perfect, distinguishable from the past imperfect *I was going to town*. Both are distinguishable from the simple past, *I went to town*. The future perfect works likewise: *I will go to town* is simple future, *I will have gone to town* is future perfect, and *I will be going to town* is future imperfect. We can also use different moods, such as the conditional to indicate hypothetical events, or the subjunctive to indicate events that are conditional on wishes or emotions, or conditions that do not apply at the present. In English, we make extensive use of auxiliaries, such as the variants of *go* and *have*, to indicate tense, but we also attach morphemes (units of meaning) to the verbs themselves, such as -*ed* to indicate the past and -*ing* to indicate ongoing action. Anyone who has learned a language such as Latin will know that verbs in that language are awash with different morphemes that have to do with time and its uncertainties.

In some languages, though, the verbs are without tense, and time is marked in other ways. Chinese, for example, has no tense, but the time of an event can be indicated by adverbs, such as *tomorrow*, and what are called aspectual markers, such as the word *before* in a sentence that might be roughly rendered as *He break his leg before*.[10] In English, too, we can use adverbs, such as *yesterday* and *tomorrow*, and other markers to specify time more precisely. These include clock times. The first aircraft to hit the Twin Towers in New York City struck at 8:46 a.m. on 11 September 2001. In our increasingly frantic lives, the precise time of events, past and future, is of great importance.[11]

Of course, referring to episodes at different points in time generally involves different places—what I did yesterday need not have taken place in where I am today or will be tomorrow. So long as communication involves only the present, place can be communicated largely by pointing or by the mere fact of shared location, but reference to places at other times itself requires an extensive symbolic vocabulary. So it is that we have names for countries, states, provinces, cities, towns, streets, houses, rooms. There are also landmarks such as mountains, lakes, rivers, forests, specific buildings such as museums, town halls, theaters, and the like. Terms to locate places relative to other places and landmarks are also needed, and include the compass directions *north*, *south*, *east*, and *west*, and relative terms such as *left* and *right*, *near* and *far*, *above* and *below*. Precision is achieved by the use of measured distance, whether in *miles*, *kilometers*, *inches*, *millimeters*. Or *light years*, which also incorporate the sense of time.

Language does not merely allow people to share episodes, it can also transport them mentally to different times and places. This requires a distinction between what have been called *reference time*, or the time referred to in the narrative, and *speech time*, the time at which the speaker is speaking.[12] Both of these may differ from *present time* if one is recalling a story told on a previous occasion. Each of these times, moreover, may locate both listener and narrator in different places. You may recall being told yesterday of X's adventures in China a year ago, or of Y's plans for her daughter's wedding. The complex set of mental perspectives requires multiple recursive embedding, which people accomplish with ease. Our lives are replete with devices to take us into different worlds at different times, whether books, plays, movies, or soap operas.

Since the understanding of space long preceded mental time travel, it is perhaps not surprising that terms relating to time have largely piggybacked on spatial terms. American Sign Language (ASL) makes use of an implicit time-line from the back of the body to the front, with the future in front and the past behind. Even in spoken languages, there is typically an implicit spatial dimension underlying our sense of time. As in ASL, English refers to the past as behind us and the future as in front, and most of our spatial pronouns, such as *about, after, around, before, by, in, near, far, toward, with*, and so on, refer to time as well as space. In the language of the Aymara, residents of the Andes, time runs the other way, with the past in front and the future behind; thus the expression for *last year* is *nayra mara*, literally *front year*, and the expression for a *future day* is *quipüru*, literally *behind day*.[13] Their gestures about events in time conform to this arrangement. In Chinese, time is represented vertically, and travels downwards, so that *the month above* means last month.[14]

The importance of time in the shaping of language is nicely illustrated by the Pirahã, the Amazonian tribe in Brazil we encountered in chapter 2. Their language is almost devoid of tense and time markers. Verbs are merely marked to indicate whether an action is within the speaker's immediate experience ("proximate") or not ("remote'), but there is otherwise no tense. There are a few words indicating different times, such as those roughly translatable as *another day, now, already, night, low water, high water, full moon, noon, sunset*, and a few others.[15] The limited ability to refer to other points in time is mirrored by their actual experience of time. As we saw in chapter 2, they have no fiction or myths.[16] Their kinship system is among the simplest ever recorded, and once they die, relatives are forgotten. Everett could find no individual who could name his or her great-grandparents, and very few who could name all four grandparents. Among the events that feature in their lives are the arrival and departure of boats on the river, often carrying traders. These events seem to exist in the present without passing into memory. Everett writes as follows:

> Pirahã's excitement at seeing a canoe go round a river bend is hard to describe; they see this almost as travelling into another dimension. It is interesting, in the light of the postulated cultural constraint on grammar, that there is an important term for crossing the border between experience and nonexperience.[17]

Benjamin Lee Whorf argued that examples like this reflected the influence of language on thought itself, an idea that impressed the linguist Edward Sapir and has become known as the Sapir-Whorf hypothesis. Whorf was especially interested in Native American languages, and noted that the Hopi language "is seen to contain no words, grammatical forms, constructions or expressions that refer directly to what we call 'time,' or to past, present or future."[18] It turns out that Whorf was wrong about this, as later work by Ekkehart Malotki showed the Hopi to have multiple ways of talking about time, including the use of tense.[19] It is in any case more in keeping with this chapter to suppose that language reflects thought, rather than vice versa. That is, the Pirahã have only limited ways to talk about time because they live largely in the present.[20]

As we saw in chapter 2, there are other ways in which Pirahã language is limited. It has no numbers or system of counting, no color terms, and may even be said to lack verbs, in the sense of a verb as a linguistic class; the Pirahã learn verbs one by one as individual entities. As we saw in chapter 4, the Pirahã also make do with fewer phonemes than any other language group.

The example of the Pirahã adds to the evidence that grammatical language is indeed a consequence of the understanding of time, and the ability to travel mentally in time. Without the burden of time, language may well be free to develop in other directions. For example, Everett notes that the Pirahã communicate almost as much by singing, whistling, and humming, and have very rich prosody, with a five-way distinction between syllable types. And they can also lie and tell jokes, two indispensable human attributes.

They have much more to communicate about than boring rounds of golf.

Generativity

The most distinctive property of language is that it is generative. We can both construct and understand sentences that we have never used or heard before.

A classic example comes from the British philosopher Alfred North Whitehead. In 1934 he had been seated at dinner next to the psychologist B. F. Skinner, who was trying to explain how behaviorism would change the face of psychology. In Skinner's view, sentences were learned sequences, built up from the principles of operant conditioning. Obliged to challenge this view, Whitehead uttered the sentence "No black scorpion is falling upon this table," and asked Skinner to explain the behavioral principles that might have led him to say that. It was not until the publication of *Verbal Behavior* 23 years later that Skinner attempted an answer.[21] In an appendix to that book, Skinner proposed that Whitehead was unconsciously expressing the fear that behaviorism might indeed take over, likening it to a black scorpion that he would not allow to tarnish his philosophy. Skinner's explanation is ironic, because it seems to owe more to psychoanalysis than to behaviorism, and Skinner was well known for anti-Freudian views.[22]

We now know, largely through the efforts of Noam Chomsky, that language cannot be explained in terms of learned sequences.[23] Instead, it depends on rules, as explained in chapter 2. These rules combine words in precise ways to enable us to extract meaning. As the German philosopher Gottlob Frege put it:

> The possibility of our understanding sentences that we have never heard before rests evidently on this, that we can construct the sense of a sentence out of parts that correspond to words.[24]

The combinatorial structure of sentences, I suggest, derives in large part from the combinatorial structure of episodes, and words provide the access to the components of episodes. Most of the episodes we witness, remember, or construct in our minds are combinations of the familiar. Indeed it is the combinations, rather than the individual elements, that make individual episodes distinct. In Whitehead's sentence, the notion of a black scorpion, falling, and a table are of themselves of less interest than the unusual combination of a scorpion in downward motion above the very table at which the two savants sat—and there may have been relief that this unusual event was *not* occurring. The manner in which the words describing such episodes are arranged depends on the conventions that make up grammar.

One such convention has to do with the order in which words are uttered or signed. The most basic episodes are those involving objects and actions, so the first "words" were probably nouns and verbs—an idea championed by the eighteenth-century English philologist John Horne Tooke, who regarded nouns and verbs as "necessary words."[25] The prototypical episode of someone doing something to someone or something else, then, requires one noun to be the subject, another to be a verb describing the action, and another noun to be the object of the action. How these are ordered is simply a matter of convention. In English, the convention is to place them in the order *subject verb object* (SVO). To use a well-chewed example, the sentence "Dog bites man" means something very different from "Man bites dog"; the latter is news, the former simply a personal misfortune—or perhaps triumph, from the dog's point of view. Among the world's languages, the most common word order is SOV, with the verb at the end. SOV languages range from Ainu to Yukaghir. Indeed all possible combinations seem to exist among the world's languages, although I'm told there are only four OSV languages.[26]

Since we can only utter one word at a time, word order can be critical in speech. Some languages, though, mark the roles played by different words with changes to the words themselves. In Latin, for example, the subject and object of a sentence are signaled by different inflections—changes to the end of the word—and the words can be reordered without losing the meaning. So *canis virum mordet* means *Dog bites man*, while *canem vir mordet* means *Man bites dog*, although it would be normal still to place the subject first. The Australian aboriginal language Walpiri is a more extreme example of an inflected language in which word order makes essentially no difference. Such languages are sometimes called *scrambling languages*. Chinese, by contrast, is an example of an *isolating language*, in which words are not inflected and different meanings are created by adding words or altering word order. English is closer to being an isolating language than a scrambling one.

Compared to speech, signed languages are not so constrained to present words in sequence, since they make use of spatial information as well as sequential information. Consider an action such as a cow jumping over the moon. Since the cow, the jumping, and the moon are all present in the action, we need a convention in speech to determine how to order them; English is an SVO language, and the ordering is *cow, jump, moon*: *The cow jumped over the moon*. In signing, though, one could pantomime the action by having one hand represent a cow, the other the moon, and have the cow hand make a jumping movement over the moon hand. Even signed languages, though, need conventions about order. As they evolve, signed languages tend toward segmentation and linearization; as we saw in chapter 2, children using NSL segmented the motion of a ball into two separate signs, and signed them separately. Segmentation provides for a more efficient combinatorial structure, allowing the same signs to be used in different combinations. Thus signing an action like a cow jumping over the moon tends to segment into separate signs for cow, jump, and moon, just as it does in speech. In American Sign Language the basic order is SVO, while the newly emerged Al-Sayyid Bedouin Sign Language (ABSL) is an SOV language.[27]

Of course there is more to episodes than simple if painful actions like dogs biting men, and correspondingly more to language. We need to record such things as whether it was a particular dog, or just some unfriendly or perhaps hungry animal that happened to be passing by. The definite article *the* (related to *that*) is used to specify a particular object or person, while the indefinite article *a* or *an* (related to *one*) refers to an object or person not specified or particularized. We may need also to record further detail about the dog, or the man, or perhaps even the nature or location of the bite. Many of the operations of grammaticalization, discussed in chapter 2, are driven by considerations of this sort. For example, embedded phrases may qualify the people or objects involved in an episode, as in sentences like *The dog, which had previously bitten my aunt, bit the postman.*

Once upon a Time

Episodic memories, along with combinatorial rules, allow us not only to create and communicate possible episodes in the future, but also to create fictional episodes. As a species, we are unique in telling stories. Indeed the dividing line between memory and fiction is blurred; every fictional story contains elements of memory, and many memories contain elements of fiction. Stories may be set in the past, as in historical fiction, or in the future, as in prophetic books such as George Orwell's *1984* (a date now thankfully in the past), or they may be set in some unspecified period of time. Of course stories also carry their own ribbon of time. Relative to any given point in a novel, there is a past and a future, although the author has the luxury of being able to flit back and forth between them at her whim.

Stories are adaptive because they allow us to go beyond personal experience to what might have been, or to what might be in the future. They provide a way of stretching and sharing experience so that we are better adapted to possible futures. Moreover, stories tend to become institutionalized, ensuring that shared information extends through large sections of the community, creating conformity and social cohesion. Examples include the Bible, the Koran, J. K. Rowling's *Harry Potter* stories, and popular soap operas. Even the near-universal predilection for murder mysteries may have adaptive significance. They alert us to events that might happen (but we hope they won't), and so provide scenarios that might follow if they do, and make us better prepared to deal with them. And the murderer nearly always gets caught.[28] This means that murder stories are morality tales, helping inculcate a sense of what one should and should not do. There is of course a dark side as well. By exposing the mistakes that lead to the murderer being caught, murder mysteries may help the readers themselves get away with murder.[29]

Not all fiction is murderous. Stories also allow us to understand more generally what goes on in the minds of others, so that we can better understand their motivations and actions. This is important for social understanding, helping us to achieve joint goals. But fiction also extends beyond the ordinary, including characters that transcend normal human capabilities. The Bible is but one example, and perhaps historically the most influential of all. Children's stories, in particular, are alive with talking animals, fairies, magicians, and other supernatural beings,[30] as is well illustrated by the extraordinary success of the *Harry Potter* series. Is the supernatural adaptive? Perhaps the stretching of the imagination allows us to better understand what might be possible. The evolutionary biologist David Sloan Wilson remarks that "Even massively fictitious beliefs can be adaptive, as long as they motivate behaviors that are adaptive in the real world."[31]

Stories about the supernatural morph naturally into religion. Brian Boyd points out that religious conviction derives less from doctrine than from stories.[32] Like other religious works, the Bible tells stories of such supernatural incidents as virgin birth, walking on water, or rising from the dead. As Boyd notes, evolution will favor belief in a falsehood if it motivates more adaptive behavior than does belief in a truth.[33] One falsehood[34] that is perhaps encouraged by tales of the supernatural is the notion of life after death; as suggested in the previous chapter, belief in the afterlife can alleviate the fear of death and subsequent oblivion, and the belief is no doubt fortified if expressed in stories that are widely shared. More generally, the adaptiveness of stories about the supernatural may derive from their sheer power to spread throughout a culture, engendering social cohesion. As Boyd puts it, "We may even see reluctance to believe as a challenge to group unity and as tantamount to treason."[35]

Beyond Time

The main argument of this chapter is that grammatical language evolved to enable us to communicate about events that do not take place in the here and now. We talk about episodes in the past, imagined or planned episodes in the future, or indeed purely imaginary episodes in the form of stories. Stories may extend beyond individual episodes, and involve multiple episodes that may switch back and forth in time. The unique properties of grammar, then, may have originated in the uniqueness of human mental time travel.

But the structure of language itself is not a matter of mental time travel. Words are stored in semantic memory, and only rarely or transiently in episodic memory. I have very little memory of the occasions on which I learned the meanings of the some 50,000 words that I know—although I can remember occasionally looking up obscure words that I didn't know, or that had escaped my semantic memory. The grammatical rules by which we string words together may be regarded as part of implicit rather than explicit memory, as automatic, perhaps, as riding a bicycle. Indeed so automatic are the rules of grammar that linguists have still not been able to elaborate all of them explicitly. Thus although language may have evolved, initially at least, for the communication of episodic information, it is itself a robust system embedded in the more secure vaults of semantic and implicit memory. It has taken over large areas of our memory systems, and indeed our brains.

And although language may have been initiated by the adaptive advantage of sharing our episodic experiences, we have also adapted language for a variety of other purposes. One such purpose is teaching. Indeed this book is an attempt to explain to you some of the things I believe to be true of the human mind, and contains little of my own personal experiences—although I have occasionally indulged, but not, I hope, to the point of boring you. The pedagogical use of language may have developed with the emergence of manufacturing techniques. In chapter 4 I suggested that language shifted from a predominantly manual mode to a predominantly vocal one relatively late in the evolution of our species since we shared a common ancestor with the chimpanzee, some six or seven million years ago. Indeed in chapter 12 I explore the possibility that the shift was not complete until the emergence of *Homo sapiens*, which may explain the surge of manufacture in our own species. So pedagogy began to encroach on the use of language for the telling of stories. We use language to explain how things work, or how to cook, as on popular cooking shows on television. But still we gossip.

In this chapter I have suggested, in effect, that the recursive nature of language evolved, at least in part, from the recursive nature of mental time travel. Recursive language structures need not map directly onto the recursive nature of imagined episodes. As we saw in chapter 2, the Pirahã use nonrecursive structures to recount recursive episodes. Conversely, a sentence may include an embedded phrase that has nothing to do with an episode itself, but may simply provide background information, as in a sentence like *George, who has a generous disposition, gave most of his money to his avaricious daughters*. So it goes.

Mental time travel, though, is not the only recursive function that has a bearing on the evolution of language. In the next chapter, I expand on another recursive property of the human mind that also played a critical role in the evolution of language.

PART 3

Theory of Mind

Another ingredient of human thought is the ability to understand, or infer, what is in the minds of other people. This is again recursive, since I might not only infer what another individual is thinking, but also infer that she infers what I am thinking. As I suggest in chapter 8, mind reading is not the product of extrasensory perception or psychic emanations, but is rather a mental process, dependent on common situations, shared experience, and an understanding that other minds are like our own. Mind reading is critical to human cooperation, but may also underlie some of our more deceitful practices, such as lying, stealing, and cheating. Believe me.

As I explain in chapter 9, language depends critically on theory of mind. Indeed it is itself one of the mechanisms by which we read the minds of others. Unlike animals calls, which consist of relatively fixed signals, language is intrinsically ambiguous, and meaning must be inferred not only from what a person says (or signs), but also from what one knows about that person and what one shares with her. A conversation is more a shared stream of thought than the transmission of information, and can often be carried on with a minimal use of words.

This part of the book continues the theme that language is not so much an independent biological faculty as part of a more general cognitive capacity—in this case the capacity to share our thoughts and emotions. The social nature of language is one reason that no one has yet succeeded in having meaningful conversations with a computer. Perhaps we might make an exception of ELIZA, the famous computer program that simulated Rogerian psychotherapy, which essentially operated by using stock phrases triggered by key words in the patient's utterances. Well, maybe some conversations are a bit like that, but I like to think that even psychotherapy may have evolved into something a little more meaningful.

8

Mind Reading

There are three classes of intellects: one which comprehends by itself; another which appreciates what others comprehend; and a third which neither comprehends by itself nor by the showing of others; the first is the most excellent, the second is good, and the third is useless.
—Niccolò Machiavelli, *The Prince*

Many people believe that thoughts can be transferred from one person to another by means other than the senses. This is known as telepathy. In 1882, the Society for Psychical Research was established in London to investigate telepathy and other so-called psychic phenomena, such as ghosts, trance states, levitations, mediums, and communication with the dead. Its first president was Henry Sidgwick, later professor of moral philosophy at Trinity College, Cambridge, and other distinguished members included the experimental physicist Lord Rayleigh, the philosopher Arthur Balfour, who became prime minister of England from 1902 to 1905, and Sir Arthur Conan Doyle, author of the Sherlock Holmes stories. The Society attracted the interest of famous psychologists such as Sigmund Freud and Carl Jung, and the pioneering American psychologist William James was so impressed that he established the American Society for Psychical Research shortly afterwards.

These societies, and many others dedicated to psychical research, remain active to this day. Laboratories for the study of psychic phenomena emerged in a number of universities, with Stanford University leading the way in 1911. Under the guidance of the famous psychologist William McDougall, Duke University set up an influential laboratory in 1930, and the publication of Joseph B. Rhine's book *New Frontiers of the Mind* in 1937 brought the laboratory's findings to the attention of the general public. Rhine and

McDougall coined the term "parapsychology" to refer to psychic phenomenon, and the *Journal of Parapsychology*, which flourishes still, was established in 1937. In 1983, the noted author Arthur Koestler and his wife Cynthia provided in their wills for the establishment of a chair in parapsychology at a British university. The University of Edinburgh took advantage of the opportunity, and in 1984 the chair was duly established there, with Robert Morris as its first incumbent. Morris died in 2005, and at the time of writing has not been replaced, although the Koestler Parapsychology Unit remains active.

As this brief introduction illustrates, parapsychology has been associated with some distinguished individuals and reputable universities. The tenuous nature of claims of psychic phenomena has also meant that fraudsters have been quick to cash in. Uri Geller, an Israeli-British stage performer, rose to fame in the 1970s for television shows on which he claimed to demonstrate psychic powers. He is perhaps best known for his prowess at bending spoons, apparently through the power of thought—a phenomenon that, if true, is an example of *psychokinesis*. Geller's demonstrations are easily duplicated without resort to psychic powers by stage magicians, including James Randi, who wrote a book entitled *The Magic of Uri Geller*—it was later called *The Truth about Geller*.[1] Geller's exploits were also unmasked, leaving no spoon unbent, in New Zealand by two psychologists, David Marks and Richard Kammann, who were able to repeat Geller's demonstrations on television, again without any claim to psychic powers. They too wrote a book exposing the field of psychic phenomena, and Geller in particular, entitled *The Psychology of the Psychic*.[2] These books are highly recommended, but alas do not have the selling power of books that proclaim the existence of the psychic.

Just why people should be so ready to believe in the power of the mind to transcend physical laws is unclear. Perhaps it is a kind of wish fulfillment, allowing us to believe that we can influence people at a distance or communicate with loved ones who have died. Perhaps it is an offshoot of mind-body dualism, attributed to the seventeenth-century French philosopher Réné Descartes, who proposed that the human mind was not governed solely by physical laws. Most religions depend on the idea of a spirit or god that goes beyond the merely physical. Some devotees have tried to give scientific respectability to psychic forces through labeling; parapsychologists have called it *psi* (pronounced "sigh"), which to more mainstream scientists, albeit old-fashioned ones, means "pounds per square inch," while a more ambitious term, coined by the biologist Rupert Sheldrake, is *morphic resonance*. Sheldrake is the fellow we met in chapter 3 who recently claimed that dogs can use morphic resonance to sense when their owners are coming home unexpectedly.[3]

Theory of Mind

Part of the reason that the belief in psychic phenomena persists, though, is that people are actually very good at reading minds. We do this through means that are entirely naturalistic, and there is no good reason to invoke supernatural powers or nonphysical channels of communication. The ability to understand, or at least surmise, what is happening in the minds of others is known as *theory of mind*. It is recursive, in the sense that it involves the insertion of what you believe to be someone else's state of mind into your own. Let's consider, then, some of the naturalistic ways in which we can do this.

Emotion is perhaps the simplest mental state to read. We can readily tell whether another person is happy, angry, sad, or in pain, through their facial expressions, bodily postures, or vocalizations. In his book *The Expression of the Emotions in Man and Animals*, Charles Darwin vividly described outward signs of emotions, and our ability to read the emotions of others is by no means confined to our own species. Darwin was equivocal, though, about whether this was learned or instinctive; as is usually the case, he is worth quoting in detail:

> As most of the movements of expression must have been gradually acquired, afterwards becoming instinctive, there seems to be some degree of *a priori* probability that their recognition would likewise have become instinctive. There is, at least, no greater difficulty in believing this

than in admitting that, when a female quadruped first bears young, she knows the cry of distress of her offspring, or than in admitting that many animals instinctively recognize and fear their enemies; and of both these statements there can be no reasonable doubt. It is however extremely difficult to prove that our children instinctively recognize any expression. I attended to this point in my first-born infant, who could not have learnt anything by associating with other children, and I was convinced that he understood a smile and received pleasure from seeing one, answering it by another, at much too early an age to have learnt anything by experience. When this child was about four months old, I made in his presence many odd noises and strange grimaces, and tried to look savage; but the noises, if not too loud, as well as the grimaces, were all taken as good jokes; and I attributed this at the time to their being preceded or accompanied by smiles. When five months old, he seemed to understand a compassionate expression and tone of voice. When a few days over six months old, his nurse pretended to cry, and I saw that his face instantly assumed a melancholy expression, with the corners of the mouth strongly depressed; now this child could rarely have seen any other child crying, and never a grown-up person crying, and I should doubt whether at so early an age he could have reasoned on the subject. Therefore it seems to me that an innate feeling must have told him that the pretended crying of his nurse expressed grief; and this through the instinct of sympathy excited grief in him.[4]

Whether instinctive or learned, the human ability to infer the mental states of others goes well beyond the detection of emotion. To take another simple and seemingly obvious example, we can understand what another individual can see. This is again an example of recursion, since we can insert that individual's experience into our own. It is by no means a trivial feat, since it requires the mental rotation and transformation of visual scenes to match what the other person can see, and the construction of visual scenes that are not immediately visible. For example, if you are talking to someone face-to-face, you know that she can see what is behind you, though you can't. Someone standing in a different location necessarily sees the world from a different angle, and to understand that person's view requires an act of mental rotation and translation.

To test this ability in children, the Swiss psychologist Jean Piaget developed what is known as the Three Mountains Test, in which the children looked at an arrangement of three model mountains from a particular location in the room. They were then shown photographs of the scene taken from different locations, and asked to select which one showed how the scene would look from some particular location. Piaget found that up to the age of nine or ten, children were unable to solve the problem when the photograph represented the view from a location other than their own.[5] This particular task seems to have been unusually difficult, though, and children as young as three or four have been shown to be able to solve a comparable task in which the mountains are replaced by more familiar objects.[6]

More complex still is the capacity to infer what other people *believe*, often on the basis of observation and reasoning. This is nicely illustrated by the Sally-Anne test, which is a test of children's ability to infer false beliefs. The child is shown a scene involving two dolls, one called Sally and one called Anne. Sally has a basket and Anne has a box. Sally then puts a marble in her basket and leaves the scene. While Sally is away Anne takes the marble out of the basket and puts it in her box. Sally then comes back, and the child is asked where she will look for her marble. Children under the age of four typically say she will look in the box, which is where the marble actually is. Older children will understand that Sally did not see the marble being shifted, and will correctly say that Sally will look in the basket. They understand that Sally has a false belief.

The Sally-Anne test is something of a classic in developmental psychology, but may overestimate the age at which children develop theory of mind. The test requires not only that the child remember a series of events, but also that the child understand the question put to her. A recent study suggests that babies may actually understand false belief by the age of two. Children aged 25 months were shown a movie of an actor placing a ball in a box. The actor then looks away and the ball is removed. When the actor returned, 17 of the 20 infants looked at the box where the ball had been placed, evidently expecting the actor to mistakenly look there for the ball. These young infants did seem to understand that the actor would have a false belief.[7]

Theory of mind evolved because we live complex social lives, where we deal as much with people as with objects—or we did, until computers appeared on our desks. Survival during the Pleistocene, when our forebears competed with dangerous carnivores on the African savanna, required cooperation and social intelligence. The story of the Pleistocene in shaping the human mind will be told in more detail in chapter 11, but suffice to say that reproductive success in humans is driven more by social success than by physical attributes.[8]

There may be a dark side to social intelligence, though, since some unscrupulous individuals may take advantage of the cooperative efforts of others, without themselves contributing. These individuals are known as freeloaders. In order to counteract their behavior, we have evolved ways of detecting them. Evolutionary psychologists refer to a "cheater-detection module" in the brain that enables us to detect these imposters, but they in turn have developed more sophisticated techniques to escape detection.[9] This recursive sequence of cheater detection and cheater-detection detection has led to what has been called a "cognitive arms race," perhaps first identified by the British evolutionary theorist Robert Trivers,[10] and later amplified by evolutionary psychologists.[11] The ability to take advantage of others through such recursive thinking has been termed *Machiavellian intelligence*,[12] whereby we use social strategies not merely to cooperate with our fellows, but also to outwit and deceive them. In the sixteenth century, in his celebrated work *The Prince*, Niccolò Machiavelli gave sage advice:

> [It] is useful, for example, to appear merciful, trustworthy, blameless, religious—*and to be so*—yet to be in such measure prepared in mind that if you need to be not so, you can and do change to the contrary.

The philosopher Daniel Dennett referred to mind reading as the *intentional stance*, which means that we tend to treat people as having intentional states.[13] The notion of intentional state is here used rather broadly, and not just the intention to act in a particular way. It includes other subjective states such as beliefs, desires, thoughts, hopes, fears, and so forth. According to the intentional stance, we interact with people according to what we think is going on in their minds, rather than in terms of their physical attributes—although there is a bit of that too, as I recall from my early days on the rugby field. When you meet a stranger on a dark night, your behavior may be guided partly by the intentional stance, based perhaps on facial expression, but perhaps also by what might be termed the *physical stance*, based on just how big a hunk the stranger is.

From the point of view of this book, the important aspect of theory of mind is that it is recursive. This is captured by the different orders of intentionality proposed by Dennett. Zero-order intentionality refers to actions or behaviors that imply no subjective state, as in reflex or automatic acts. First-order intentionality involves a single subjective term, as in *Alice wants Fred to go away*. Second-order intentionality would involve two such terms, as in *Ted thinks Alice wants Fred to go away*. It is at this level that theory of mind begins. And so to third order: *Alice believes that Fred thinks she wants him to go away*. Recursion kicks in once we get beyond first order, and our social life is replete with such examples. There seems some reason to believe, though, that we lose track at about the fifth or sixth order,[14] perhaps because of limited working-memory capacity rather than any intrinsic limit on recursion itself. We can perhaps just wrap our minds around propositions like: *Ted **suspects** that Alice **believes** that he does indeed **suspect** that Fred **thinks** that she **wants** him (Fred) to go away*. That's fifth order, as you can tell by counting the words in bold type. You could make it sixth order by adding *George **imagines that** . . .* at the beginning.

According to Robin Dunbar, it is through theory of mind that people may have come to know God, as it were. The notion of a God who is kind, who watches over us, who punishes, who admits us to Heaven if we are suitably virtuous, depends on the understanding that other beings—in this case a supposedly supernatural one—can have human-like thoughts and emotions. Indeed Dunbar argues that several orders of intentionality may be required, since religion is a social activity, dependent on shared beliefs. The recursive loops that are necessary run something like this: *I suppose that you think that I believe there are gods who intend to influence our futures because they understand our desires.*[15] This is fifth-order intentionality.[16] Dunbar himself must have achieved sixth-order intentionality if he supposes all this, and if you suppose that he does then you have achieved seventh-order. Call it Seventh Heaven, if you like.

If God depends on theory of mind, so too, perhaps, does the concept of self. This returns us to the opening paragraph of this book, and Descartes's famous syllogism "I think, therefore I am." Since he was appealing to his own thought about thinking, this is second-order intentionality. Of course we also understand the self to continue through time, which requires the (recursive) understanding that our consciousness also transcends the present.

People and Things

A few individuals appear to lack the ability to read minds, and Simon Baron-Cohen has argued that this deficit, which he calls *mindblindness*, underlies the condition known as autism.[17] People with this condition may be intelligent in other respects, but are peculiarly unresponsive to other people. One remarkable case is a woman called Temple Grandin, who has a Ph.D. in agricultural science and works as a teacher and researcher at Colorado State University. Clearly intelligent, she has written several books, three of which describe her own condition and the manner in which she has dealt with it.[18] Her plight was also vividly described by Oliver Sacks in his book *An Anthropologist on Mars*.[19] She has had to laboriously teach herself how people act in different circumstances, so that she knows how to act appropriately in social situations. One bonus arising from this strategy is that her habit of detailed observations of behavior has provided insights into the behavior of animals, as revealed in her most recent book, *Animals in Translation: Using the Mysteries of Autism to Decode Animal*

Behavior.[20] This book prompted a documentary, broadcast on the BBC on 8 June 2006, rather unkindly entitled "The Woman Who Thinks Like a Cow." Curiously, though, she is sensitive to emotion in others, and is herself emotionally sensitive. The very specific nature of this disability has led some evolutionary psychologists to propose that theory of mind is a *module*, independent of other aspects of the human mind.

High-functioning autism, as evident in people like Temple Grandin, is known as Asperger's syndrome. People with this condition often can pass false-belief tests, such as the Sally-Anne test, but they apparently do so only through verbal reasoning and explicit instructions about the task. As I noted earlier, normal infants seem to instinctively demonstrate an understanding of false belief well before they can demonstrate it verbally, since they will look to where an actor mistakenly believes an object to be hidden. People with Asperger's syndrome do not do this, suggesting that the spontaneous understanding of false belief is lacking.[21]

At the opposite end of the spectrum to autism, it has been suggested, lies psychosis.[22] At least some aspects of psychosis seem to reflect a hypermentalism. The more florid symptoms of schizophrenia, for example, include hallucinations, delusions, and paranoia. It's as though schizophrenics read too much into the minds of others, to the point that they think there are plots to exterminate them, or that their minds are controlled by some sinister external agency. The Scottish psychiatrist R. D. Laing[23] was particularly adept at expressing the recursive mentality that can cause social relationships to go wrong, and psychosis to develop. Here are excerpts from his aptly titled book *Knots*:

JILL: I'm upset you are upset

JACK: I'm not upset

JILL: I'm upset that you're not upset that I'm upset that you're upset.

JACK: I'm upset that you're upset that I'm not upset that you're upset that I'm upset, when I'm not.

JILL: You put me in the wrong

JACK: I am not putting you in the wrong

JILL: You put me in the wrong for thinking you put me in the wrong.

JACK: Forgive me

JILL: No

JACK: I'll never forgive you for not forgiving me

The extremes of autism and psychosis may lie on a continuum that governs normal as well as abnormal behavior. Some personality theorists write of the schizotypal personality, with a tendency toward paranoia and magical thinking, but not classifiable as psychotic. Toward the other end of the continuum, some individuals show autistic tendencies, which may include specific language impairments and obsessive-compulsive behaviors. Think of computer geeks. (Perhaps you are even one yourself, although I suppose if you are, you mightn't know it.) The continuum may also be considered one of mechanism versus mentalism. We live in a complex world of things and people, and nature seems to have provided us with sufficient flexibility to deal with both. As the evolutionary biologist William D. Hamilton put it, "There are people people and things people."[24]

Things people may tend to adopt what I have called the physical stance, in which people are regarded more as moving lumps of meat than as mental beings. Perhaps radical behaviorists, and rugby players, tend to see people in this way. Conversely, people people may tend to treat things as people. Inanimate objects, such as cars, ships, or computers, are often ascribed human-like properties. Throughout history, and perhaps human prehistory, people have personified inanimate objects, such as the stars and planets, and have bestowed human properties on nonhuman animals. Personified animals dwell especially comfortably in children's books, as we saw in chapter 3. And then there's God, encountered earlier, who appears to be personification without material substance—the ultimate triumph, perhaps, of theory of mind.

Sex and the Brain

To a degree at least, the autistic-psychotic spectrum seems to underlie differences between men and women. Simon Baron-Cohen has described autism as an extreme of male behavior, and men in general seem more concerned with things than with people[25]—and indeed may tend to treat women as things rather than as people.

Despite the efforts of feminists, young boys seem to prefer toy tractors or spaceships, while young girls go for dolls and their mothers' makeup. The positive symptoms of schizophrenia are more common in females than in males, although negative symptoms and schizophrenia as a whole are somewhat more common in males. And women tend to be more religious—as we have seen, this can be considered a rather complex manifestation of theory of mind. Students of psychology, the science of the mind, are predominantly female—at least in the departments of psychology I am familiar with—although in the old days of behaviorism there were more men about. Men may be more drawn to the sciences of the physical world.

The biological basis of the autistic-psychotic spectrum may depend more on maternal and paternal genes than on the biological sex of the offspring. The battle of the sexes begins in the womb, through the phenomenon of *imprinting*. Chromosomes come in pairs, one from the mother and one from the father, and imprinting means that one or other can dominate. The parents have different interests in the fate of the offspring, and this can be expressed in the relative influence of maternal and paternal genes. In mammalian species, the only obligatory contribution of the male to the offspring is the sperm, and the father relies primarily on his genes to influence the offspring to behave in ways that support his biological interest. Paternal genes should therefore favor self-interested behavior in the offspring, drawing on the mother's resources and preventing her from using resources on offspring that might have been sired by other fathers. The mother, on the other hand, has continuing investment in the child both before birth, in terms of nutrient from her own body, and after birth, in terms of breast milk and care-giving. Maternal genes should therefore operate to conserve her resources, favoring sociability and educability[26]—nice kids, who go to school and do what they're told.

Maternal genes are expressed most strongly in the cortex, representing theory of mind, language, and social competence, whereas paternal genes tend to be expressed more in the limbic system, which deals with resource-demanding basic drives, such as aggression, appetites, and emotion. Autism, then, can be regarded as the extreme expression of paternal genes, schizophrenia as the extreme expression of maternal genes. Many of the characteristics linked to the autistic and psychotic spectra are physical, and can be readily understood in terms of the struggle for maternal resources. The autistic spectrum is associated with overgrowth of the placenta, larger brain size, higher levels of growth factors, and the psychotic spectrum with placental undergrowth, smaller brain size, and slow growth.[27]

The relative dominance of maternal and paternal genes should not be confused with the effects of the sex chromosomes themselves. Not all males are aggressive, overgrown hunks who can't read books and don't eat quiche, and not all females are polite blue-stocking vegetarians who devour the novels of Jane Austen. Nevertheless there does seem to be a correlation between imprinting and biological sex, with males leaning toward the autistic end of the continuum and females toward the schizophrenic end. This may arise because the X chromosome, along with chromosomes 2, 3, 5, 7, 9, 10, 15, 16, and 17, is one of the chromosomes subject to differential imprinting,[28] although presumably only in females since males receive only one copy of the X chromosome through the mother. Given the multitude of chromosomes involved, the effects of imprinting are likely to be diverse, and not exclusively associated with one or other sex.

Imprinting may have played a major role in human evolution. One suggestion is that evolution of the human brain was driven by the progressive influence of maternal genes, leading to expansion of the neocortex and the emergence of recursive cognition, including language and theory of mind. The persisting influence of paternal genes, though, may have preserved the overall balance between people people and things people, while also permitting a degree of difference. Simon Baron-Cohen has suggested that the dimension can also be understood along an axis of empathizers versus systematizers.[29] People people tend to empathize with others, through adopting the intentional stance and the ability to take the perspective of others. Things people may excel at synthesizing, through obsessive attention to detail and compulsive extraction of rules.[30]

Although one might suppose that a balance of maternal and paternal genes is optimal, it is perhaps the departures from genetic balance that add innovation and creativity to our lives. The pioneering British psychiatrist Henry Maudsley noted that insanity was often accompanied by creative genius. One example was the eighteenth-century English essayist Charles Lamb, who was prone to periods of mental illness. Maudsley commented as follows

> His [Lamb's] case, too, may show that the insane temperament is compatible with, and indeed not seldom coexists with, considerable genius.[31]

Recent evidence suggests that a particular polymorphism on a gene known to be related to the risk of psychosis is also related to creativity in people with high intellectual achievement.[32]

The tendency to schizophrenia or bipolar disorder may underlie creativity in the arts, as exemplified by musicians such as Béla Bartók, Ludwig van Beethoven, Maurice Ravel, or Peter Warlock, artists such as Amedeo Clemente Modigliani, Maurice Utrillo, or Vincent van Gogh, and writers such as Jack Kerouac, D. H. Lawrence, Eugene O'Neill, or Marcel Proust. The esteemed mathematician John Forbes Nash, subject of the Hollywood movie *A Beautiful Mind*, is another example. The late David Horrobin went so far as to argue that people with schizophrenia were regarded as the visionaries who shaped human destiny itself, and it was only with the Industrial Revolution, and a change in diet, that schizophrenics were seen as mentally ill.[33]

The tendency to autism, especially high-functioning autism (Asperger's syndrome) may lead to genius of a different kind. This includes savants, individuals with extraordinary ability in some narrow domain, such as calculation, music, or even language, but in other respects subnormal. In chapter 5 we encountered the savant and calculation wizard Kim Peek. Savantism may arise from the narrow focusing on one activity at the expense of others. The tendency to autism, though, may also underlie high-level achievement in mathematics, physical sciences, or engineering, coupled with obsessional thinking. Isaac Newton may be a prototypical example. Another is the Nobel Prize–winning physicist Paul Dirac, regarded by some as being in the same league as Albert Einstein, but also described by his fellow physicist Niels Bohr as "the strangest man."[34] Things people may have their uses after all, since they help us understand an increasingly complex physical world. And we should not deny them the capacity to think recursively.

Recursion, then, is not the exclusive preserve of social interaction. Our mechanical world is as recursively complex as is the social world. There are wheels within wheels, engines within engines, computers within computers. Cities are containers built of containers within containers, going right down, I suppose, to handbags and pockets within our clothing. Recursive routines are a commonplace in computer programming, and it is mathematics that gives us the clearest idea of what recursion is all about. But recursion may well have stemmed from runaway theory of mind, and been later released into the mechanical world. I explore this further in chapter 12.

Do Animals Have Theory of Mind?

As Darwin recognized in the extract quoted earlier, there is little doubt that other species can read emotion in others—life often depends on it. In all animals in which reproduction depends on feeding, cleaning, warming, and protecting the young, parents are sensitive to the emotional state of the infant, and act to relieve distress or satisfy hunger or thirst. Many animals also respond with sympathy to perceived distress in others. In one study, for example, mice perceiving pain in other mice intensified their own reaction to pain,[35] and in another study monkeys refused to pull a chain to receive food if doing so caused a shock to be delivered to another monkey.[36] Frans de Waal documents these and other examples, and notes that chimpanzees, but not monkeys, often give consolation to others in distress. His ready camera records a juvenile chimpanzee putting a consoling arm around a screaming adult who has just been defeated in a fight.[37]

More problematic, though, are the more cognitive aspects of theory of mind, such as knowing what another individual sees or believes. It is natural to look first in our closest primate ancestor, the chimpanzee. The question was first put by David Premack and Guy Woodruff in 1978, in a classic paper entitled "Does the Chimpanzee Have a Theory of Mind?" One of their techniques was to show videos of a human grappling with some problem, and then offer the chimpanzee Sarah a choice of photographs, one of which depicted a solution to the problem. One such test showed a woman trying to escape from a locked cage, and one of the photos showed a key, while others showed objects irrelevant to the task. Sarah performed quite well at choosing the appropriate photo, although as Premack and Woodruff recognized, this need not show that she appreciated what was going on in the mind of the person depicted. For example, the key might have been selected through simple association with the cage.

One researcher who has taken up Premack and Woodruff's challenge is Daniel Povinelli, who has tried further tests to determine whether a chimpanzee can understand what is going on in the mind of a person.[38] The results have been largely negative. Chimpanzees readily approach humans to beg for food, and this provided an opportunity to check whether in so doing they are influenced by whether the person can see or not. But when offered the choice of two individuals to beg from, one with a blindfold over her eyes, the animals did not systematically choose the one who could see. The same was true when one of the people had a bucket over her head, or covered her eyes with her hands. Only when one of the people was actually facing the other way did the chimpanzees easily choose the one facing toward themselves. Young children, on the other hand, quickly recognize that they should approach the person who can see them, and understand that this depends on the eyes. The failure of the chimpanzee to appreciate this does not arise from failure to observe the eyes, since they readily follow the gaze of a person confronting them. Chimpanzees may eventually choose the person who can see them, but the behavior is more simply explained by associative learning, and not on the understanding that eyes are for seeing.

Another test depends on the chimpanzee's apparent understanding of pointing. If a person sits in front of a chimpanzee and points to one of two boxes to left or right, the chimpanzee understands readily enough that if it wants food, it should go to the box that the person is pointing to. But the choice breaks down if the person points from some distance away, and is systematically reversed if the person sits closer to the box that does not contain the food and points to the other one. It seems that chimpanzees respond on the basis of how close the pointing hand is to the box containing the food, and not on the basis of where the hand is actually pointing.

Yet chimpanzees do follow eye gaze,[39] just as humans do, and this may suggest that they have at least some understanding that others can see. Povinelli argues, though, that behaviors like following eye gaze have the same instinctive basis in humans as in other primates, but that we "reinterpret" these behaviors as being more sophisticated than they really are.[40] For example, we may spontaneously follow the gaze of someone who seems to be gazing at something in the sky without going through an intellectual (and presumably conscious) exercise along the lines of *That fellow must be able to see something up there that's interesting.* Gaze following may simply be an adaptive response that alerts other animals to danger or reward, but we humans have intellectualized it, often after the fact. Remember too that even infants gaze in anticipation of where an actor will falsely believe an object to be hidden, well before they can intellectualize the reason for their own behavior.

Other work suggests, though, that the work of Povinelli and his colleagues underestimates the social intelligence of the chimpanzee. Brian Hare and colleagues have shown that a chimpanzee will approach food when a more dominant chimpanzee cannot see the food, but will be reluctant to do so when they can see that the food is visible to the dominant chimpanzee.[41] Chimpanzees also seem able to pass the equivalent of the Sally-Anne test, referred to earlier. Subordinate chimpanzees will retrieve hidden food if a dominant chimpanzee was not watching while the food was hidden, or if the food was moved to another location while the dominant chimp wasn't watching. That is, the subordinate chimps appear to have knowledge of what the dominant knows. The subordinates will also retrieve food if a dominant chimp watched it being hidden, but is then replaced by another dominant chimp who hadn't watched, suggesting the subordinates can keep track of who knows what. The subordinates failed on another test, though, in which a dominant chimp watched one piece of food being hidden but did not watch another piece being hidden; they failed to consistently choose the food that the dominant chimp hadn't watched.[42] Even so, the chimps in this study seemed to have at least some knowledge of what another chimp knew.

Chimps have also been observed to hide things from each other. In one study, chimps contested access to food with a human experimenter, and chose to approach the food via a route hidden from the experimenter's view.[43] Sometimes this route was a circuitous one. Brian Hare has also shown that dogs, unlike Povinelli's chimps, can choose food sources according to where either a person or another dog is looking or pointing.[44] It is not clear why Hare's chimpanzees and dogs seem more blessed with theory of mind than were the chimpanzees studied by Povinelli.[45]

A difficulty in these studies is that they need not imply an intentional stance.[46] The animals might simply be responding on the basis of learned cues, without any sense that the action is based on an understanding of what is going on in another animal's mind—much as people with Asperger's syndrome can solve theory-of-mind problems without true understanding. If you are a young chimpanzee, the mere sight of a dominant male in the vicinity may act as a simple, learned signal not to behave in certain ways, just as a human child may wait until his mother's back is turned before punching his younger sister. But a measure of intentionality may be introduced when she later cries even if not hit, so that her older brother will be punished. In order to infer genuine theory of mind, then, we need some evidence that an act has been improvised, rather than based on trial-and-error learning. Richard Byrne, in his book *The Thinking Ape*, gives an example that may qualify:

> A young baboon, coming across an adult baboon who had just exhumed a root to eat, lets out a scream. This scream alerted the young baboon's mother, who outranked the adult in status. The mother rushed over, took in the scene of the adult, the root and the seemingly distressed young animal and chased the adult off. Meanwhile the young baboon ate the root.

Behaviors like this appear to be intentionally deceptive, and are referred to as *tactical deception*. Deception itself is widespread in nature, whether in the camouflage of a butterfly wing or the uncanny ability of the Australian lyre bird to imitate the sounds of other species—as mentioned in chapter 3, this includes the sound of a beer can being opened. Tactical deception, however, is that in which the deception is based on an appreciation of what the deceived animal is actually thinking, or what it can see—that is, it implies an intentional stance. The most obvious form of tactical deception in humans is telling lies, which we do in the hope that the recipient will believe what we say, but in nonverbal animals we must seek evidence from their actions. Andrew Whiten and Richard Byrne asked primate researchers for anecdotes from their field studies involving deceptive behavior, and carefully screened out cases in which the animals might have learned the behavior through trial and error.[47] They concluded that only the four species of ape occasionally showed evidence of having deceived on the basis of an understanding of what the deceived animal could see or know. Even so, there were relatively few instances—out of 253 observations, only 18 met Whiten and Byrne's criteria—12 from common chimpanzees and three each from bonobos, gorillas, and orangutans. These numbers are perhaps too small to convincingly demonstrate that great apes can truly "read the minds" of others.

It is sometimes suggested that there is a sort of intermediate between the physical stance and the intentional stance—a biological stance, if you will. Thus young children and apes may be able to attribute goal-directedness and self-generated motion to others, without being able to attribute full-blown intentional states, such as believing, desiring, seeing, or remembering.[48] This does not seem to involve recursion; attributing self-generated motion to another creature carries no implication that the other creature can also attribute self-generated motion. There seems no reason at present to suppose that the great apes can advance much beyond this level of attribution.[49]

Self-Awareness in Other Species?

What, then, of the ability to know one's own mind? Could a chimpanzee emulate Descartes and infer its own existence? One proposed way to answer this question is the so-called mirror test, in which a mark is applied to the animal's face, and the animal is then allowed to see itself in a mirror. The question is whether the animal will understand that the mark is on its own face, rather than on that of some stranger in the mirror. It is common practice to anesthetize the animal before applying the mark, so the animal is unaware of the mark being placed. Chimpanzees with prior experience of mirrors reach up and touch the marks on their faces, while those without experience with mirrors react as though there is another chimpanzee in the mirror, as do rhesus monkeys even after many years of experience.[50] It appears that the great apes pass the test, although not consistently, whereas lesser apes and monkeys do not, and there is weak evidence that an elephant and a dolphin may also pass. Children typically pass the test by two years of age.[51]

But even this test need not imply more than a physical stance; that is, the animal may recognize itself as a physical object, without any necessary understanding that that object has desires, beliefs, emotions, and the like. Even we humans may use mirrors to reflect ourselves as objects, rather than as sentient beings—a face to be shaved, or rendered more beautiful. The reflection in the mirror, after all, is not someone with whom one can have a meaningful conversation.

Where Are We?

The question of whether chimps and other species have theory of mind remains highly contested. Some authors continue to reject all claims of theory of mind in nonhuman species, and indeed it may always be possible to explain apparently human-like intentional behavior in terms of behavioristic principles. Derek C. Penn, Keith J. Holyoak, and Daniel C. Povinelli, in an article headed "Darwin's Mistake," conclude as follows:

> Our most important claim . . . is simply that whatever "good trick" . . . was responsible for the advent of human beings' ability to reinterpret the world in a symbolic-relational fashion, it evolved in only one lineage—ours. Nonhuman animals didn't (and still don't) get it.[52]

In this view, activities in nonhuman animals that seem to imply human-like cognition, such as theory of mind, are reinterpreted in symbolic terms. They dub their theory the "relational reinterpretation" theory.[53]

There remains the persisting question, though, of what explains the discontinuity. Despite referring to "Darwin's mistake," Penn and colleagues continue to use the term "evolved" in the above extract, but reference to a "good trick" gets us little further than reference to God or some miraculous mutation. Moreover, the appeal to "reinterpretation" is unparsimonious, implying an unnecessary step between what we do and how we interpret it. If a chimpanzee hides from prying human eyes, there seems no good reason to suppose that its behavior is any different in principle from that of the naughty child who hides from an angry parent. It smacks of arrogance to dismiss human-like behavior in animals on the grounds that humans have somehow reinterpreted such behavior in our own species to take it beyond the ken of the merely animal.

My own tentative view is that chimpanzees may indeed have some capacity to discern what other individuals can feel, see, and perhaps know. This is first-order recursion, at best. What they may lack, though, is the extension to higher-order recursion—my knowledge that another individual knows what I can see, know, or feel, or even that the other knows that I know what she's thinking. Whether or not this can explain the full repertoire of distinctively human activities, it does suggest that the difference is more one of degree than one of kind—that small step that proved such a giant leap for humankind.

Runaway theory of mind might be attributed to the highly social lives that we humans live, whether in the nuclear family, the office, the hunting party, or the rugby scrum. In such settings we must constantly monitor our actions in accord with how others think, believe, and feel. The conditions that led to the evolution of complex social behavior are discussed further in chapter 11.

Let's now return to language, since it too, in its predatory way, depends on theory of mind, and provides further insight into its recursive structure.

9

Language and Mind

Political language—and with variations this is true of all political parties, from Conservatives to Anarchists—is designed to make lies sound truthful and murder respectable, and to give an appearance of solidity to pure wind.

—George Orwell, *Politics and the English Language* (1946)

In his novel *Nineteen Eighty-four*, Orwell painted a grim picture of a future in which the ultimate technology for thought control was the language Newspeak, which could render impossible all modes of thought other than those required by Ingsoc (English Socialism). We have struggled past 1984, but political life, at least, is still replete with euphemisms designed to make us think differently. Thus *collateral damage* is a way of referring to the killing of innocent people during war, *underprivileged* means poor, *special* means handicapped, *liquidate* means murder. An extreme movement known as General Semantics was established in 1933 by Count Alfred Kozybski, an engineer, and popularized in bestsellers such as Stuart Chase's *Tyranny of Words*, and Samuel Ichiye Hayakawa's *Language in Thought and Action*. Hayakawa was later president of San Francisco State College, and gained notoriety for stamping out student protest. According to General Semantics, human folly is created by semantic damage brought about by the structure of language.

The relation between language and thought is one of the most contentious issues in the history of philosophy. As we saw in chapter 2, Chomsky's concept of I-language—the common language underlying E-languages—is essentially the language of thought. This is encapsulated also in the so-called language of thought hypothesis proposed by the philosopher Jerry Fodor, who argued that virtually all of the concepts underlying words are innate.[1] Steven Pinker refers to this as the theory that "we are born with some 50,000 concepts," based on the number of words in the typical English speaker's vocabulary.[2] Of course the actual words we use will depend on the linguistic environment a person is exposed to, but it is as though we have been already supplied with all the meanings we shall ever want, and all we need do is discover the verbal labels. It is of course difficult to believe that a person alive during the Renaissance could have been supplied with the meaning of the word *helicopter*, although one might perhaps make an exception of Leonardo da Vinci, who indeed developed the idea of such a contraption.

The idea of a strong connection between language and thought implies that nonhuman animals are incapable of thinking as we humans do, an idea defended by the psychologist Clive Wynn in his 2004 book *Do Animals Think?* If nothing else, the idea of dumb animals—dumb in both senses of the word—is a source of comfort, because it helps justify our appalling treatment of our fellow creatures, as I noted in the preface to this book. Not only do we treat them badly, but we also use them for insults, as when we characterize people as brutish, beastly, swinish, mousey, mulish, catty, hawkish, foxy, bullish, and lazy cows.

Most such expressions are unfair to the animal in question. Consider, for example, the phrase *pissed as a newt*—as far as I know, newts are not known for inebriation.[3] There are of course counterinfluences, such as the animal rights movement, which in its extreme form seeks to reverse the exploitation, threatening to harm or even kill people who exploit animals, with the aim, I suppose, of redressing the balance. Less extreme is animal welfare legislation that seeks to minimize harm to animals. Ironically, perhaps, it was in Nazi Germany that hunting foxes with hounds was first banned, on the orders of Hermann Göring in 1934. In Britain, the Hunting Act of 2004 states that "a person commits an offence if he hunts a wild mammal with a dog, unless his hunting is exempt."[4]

A modicum of thought, though, should convince us that other animals have at least a modicum of thought. The companionship of cats and dogs depends on behaviors that are not merely reflex. The behaviorists even supposed that the laws governing human action could be understood through experiments on rats or pigeons. Of course the behaviorists preferred not to speak of consciousness or mental events, so that even human activity was described in terms of behaviors, not thoughts, but the idea of a continuity between animals and humans was paramount. As John B. Watson, the founder of behaviorism, put it, "Behaviorism . . . recognizes no dividing line between men and brutes."[5] In chapter 3 I described how starlings were able to solve problems in parsing sequences that their experimenters believed to require recursive processing. By adopting a simpler but still clever strategy, the starlings may be said to have out-thought their human keepers. In chapter 3, too, I noted Wolfgang Köhler's classic experiments on insight in chimpanzees.

Perhaps the most eloquent advocate for the recognition of animal thought and consciousness was the late Donald R. Griffin, whose 1976 book *The Question of Animal Awareness* claimed that animal communication offered "a possible window on the animal mind." Griffin continued to write on this topic, and his most recent book, *Animal Minds: Beyond Cognition to Consciousness* (2001) opens with the following anecdote:

> A hungry chimpanzee walking through his native rain forest comes upon a large *Panda oleosa* nut lying on the ground under one of the widely scattered Panda trees. He knows that these nuts are much too hard to open with his hands or teeth and although he can use pieces of wood or relatively soft rocks to batter open the more abundant *Coula edulis* nuts, these tough Panda nuts can only be cracked by pounding them with a very hard piece of rock. Very few stones are available in the rain forest, but he walks 80 meters straight to another tree where several days ago he had cracked open a Panda nut with a large chunk of granite. He carries this rock back to the nut he has just found, places it in a crotch between two buttress roots, and cracks it open with a few well-aimed blows.[6]

Despite such anecdotes, it is difficult to disagree with Clive Wynn's assertion that animals do not think as we humans do. Of course they can do lots of things that we humans cannot do; examples include echolocation in bats and global navigation in migrating birds, and our clumsy machines that fly pale beside the effortless grace of a bird on the wing. Birds that cache food perform outstanding feats of memory, remembering thousands of different locations. A recent study shows a remarkable performance of chimpanzees in remembering the locations of briefly displayed numerals; one chimp outperformed a group of university students on this task.[7] So what is missing?

The evidence reviewed so far in this book indicates that nonhuman animals, even chimpanzees, are essentially incapable of theory of mind, mental time travel, or language—or are at best capable of these capacities only sporadically, and only in rudimentary fashion. I argued in chapter 7 that grammatical language depends on mental time travel, and the adaptive advantage of being able to share episodes. The final ingredient essential to the evolution of language may have been that other recursive function discussed in chapter 8—theory of mind.

Language and Theory of Mind

There is a well-known joke about a man who meets a business rival at a train station and asks where he is going. The business rival replies he is going to Minsk. The first man then says, "You're telling me you're going to Minsk because you want me to think you're going to Pinsk. But I happen to know that you are going to Minsk, so why are you lying to me?"

Commenting on this conversation, Steven Pinker writes:

> If a speaker and a listener were ever to work through the tacit propositions that underlie their conversation, the depth of the recursively embedded mental states would be dizzying.[8]

This is an exaggerated example, but in normal conversation we hardly ever spell out exactly what we mean. Instead, we depend on shared assumptions about what is going on in each other's minds. You might meet a colleague at work one morning and say, *Hey, that was a great game.* You know that she watched the rugby match last night, and would have appreciated its quality. You know too that she knows that you are referring to that game, and not to some other event, and you know that she knows that you know this. If you are speaking to a less familiar colleague, though, you might offer more information: *Did you watch the rugby last night? Great game, wasn't it?* Or, to a bemused visitor from the United States: *Rugby, you know, is our national passion. There was a game shown on TV last night, and if you happened to watch it you would have seen rugby at its best.* We therefore calibrate our conversational remarks according to theory of mind—our often implicit assumptions about what the recipient understands and knows.

Sometimes, of course, language is directed at a broader audience, and even as I write this paragraph I have in mind the minds of the audience I hope to reach. That audience, of course, includes you. I assume for a start that you understand English, and that your understanding of the world is much the same as mine, so that the various illustrations I use will make sense. I nevertheless try to spell out the arguments more than I might do if communicating with a close colleague in the same field—such colleagues won't read the book anyway, since they know me and the stuff I write all too well. The mental work underlying our utterances, whether in casual conversation or in delivering a formal lecture, is recognized by the discipline known as cognitive linguistics. One of its foremost proponents, Gilles Fauconnier, writes that "when we engage in any language activity, we draw unconsciously on vast cognitive and cultural resources, call up models and frames, set up multiple connections, coordinate large arrays of information, and engage in creative mappings, transfers, and elaborations."[9]

The essential role of theory of mind in language can be credited to the philosopher Paul Grice, who held that true language requires that the speaker has the intention of changing belief in the mind of the listener by means of the recognition of that intention. (How's that for a recursive sentence?). Grice was much concerned with the complex reasoning that goes into decoding what a given sentence might mean. He gives an example of what might underlie an utterance, P, in relation to a certain unstated thought, Q:

He said that P; he could not have done this unless he thought that Q; he knows (and knows that I know that he knows) that I will realize that it is necessary to suppose that Q; he has done nothing to stop me thinking that Q; so he intends me to think, or is at least willing for me to think, that Q.[10]

Unraveling this recursive tangle, though, would seem extremely complex, and at odds with the apparent ease with which people converse. The various thought processes and intentions underlying such statements are known as *implicatures*, and the manner in which speakers and listeners determine implicatures is one of the goals of the branch of linguistics known as *pragmatics*. Dan Sperber and Deirdre Wilson have argued that the implicatures depend on a specialized theory-of-mind *module*, in the sense proposed by Jerry Fodor, and briefly discussed in chapter 1. Modules are assumed to be innate, and to operate automatically, so one might suppose that they carry out operations of a complexity similar to those that, say, allow us to maintain balance while walking. Even so, simply declaring a function to depend on an innate module doesn't tell us how it actually works.

Sperber and Wilson suggest that there are submodules that help narrow down the alternative meanings, and so reduce the computational demand. For instance, we have a built-in sensitivity to where others are looking, and this can establish a common focus of attention. A statement such as *That's really weird* can then be quickly understood to refer to any object at that focus. More generally, Sperber and Wilson suggest that we continually maximize the *relevance* of available inputs, whether from the outside world or from memory, which can include knowledge of the memories and cultural habits of the person we are conversing with. This immediately narrows the possible interpretations of utterances, and may allow a conversation to proceed with minimal unraveling of the possible implicatures.

Sperber and Wilson provide an illustration of how this might work:

Suppose that Peter and Mary are walking in the park. They are engaged in conversation; there are trees, flowers, birds, and people all around them. Still, when Peter sees their acquaintance John in a group of people coming towards them, he correctly predicts that Mary will notice John, remember that he moved to Australia three months earlier, infer that there must be some reason he is back in London, and conclude that it would be appropriate to ask him about this. Peter predicts Mary's train of thought so easily, and in such a familiar way, that it is not always appreciated how remarkable this is from a cognitive point of view.[11]

So-called relevance theory, as developed more fully elsewhere by Sperber and Wilson,[12] suggests that our minds are focused, moment to moment, in such a way as to tune our thought processes to what is most relevant, and so reduce the linguistic demand. In conversation, we may be impervious to all but the topic under discussion, and oblivious to other events in the environment. This immediately disambiguates what might otherwise be multiply ambiguous statements. Any unexpected happening would likewise shift the shared mental processing, so participants in the conversation could immediately reach a mutual understanding of what to talk about. Language, then, is a meeting of minds, and conversation often does little more than float on the surface of shared streams of thought

At least in conversation, then, language requires what Dennett called the intentional stance, in which we treat people as mental rather than physical beings. We make implicit assumptions about the extent to which a listener is tuned into the same stream of thought, and adjust our language accordingly. Sometimes, of course, we err, as in lectures when students manifestly have no idea what I'm talking about. One might swear at a dog, or a car that won't start, and even tell stories to them. A more extreme case is illustrated by the following lyric from the 1951 Lerner and Loewe musical *Paint Your Wagon*:

> I talk to the trees
> But they won't listen to me
> I talk to the stars
> But they never hear me
> The breeze hasn't time

> To stop and hear what I say
> I talk to them all in vain.

In varying degrees, natural language is an exercise in minimalism,[13] providing just enough information to direct mutual trains of thought. This is reflected in two of Grice's famous Maxims:[14]

- Make your contribution as informative as is required for the current purposes of the exchange.
- Do not make your contribution more informative than is required.

Minimalism is exploited in the plays of the Irish author Samuel Beckett, and conveyed in his poem *What is the word?* which ends thus:

> glimpse -
> seem to glimpse -
> need to seem to glimpse -
> afaint afar away over there what -
> folly for to need to seem to glimpse afaint afar away over there
>   what -
> what -
> what is the word -
> what is the word

We saw in the previous chapter that people with autism suffer deficits in theory of mind. If theory of mind is critical to normal language, we should expect them also to have difficulties with language. And they do.[15] Even Temple Grandin, the high-functioning autistic woman we encountered in the previous chapter, struggles with language, despite the fact that she has written books and teaches in a university. She didn't begin to speak until the age of three and a half, and then used words primarily to refer to things rather than people. In the words of W. D. Hamilton, she is a prototypical things person rather than a person person. She was teased at school for the mechanical way she talked, and was incapable of gossip or chitchat. Perhaps it's not unfair to say she learned language in much the same way as an animal might learn to perform tricks, and not as a way to share information.

Irony and Metaphor

> As philosophers claim that no true philosophy is possible without doubt, by the same token, one may claim that no authentic human life is possible without irony.
>
> —Søren Kierkegaard, *The Concept of Irony* (1841)

Irony provides an excellent example of the role of theory of mind.[16] It refers to those occasions on which we say the precise opposite of what we mean, with the understanding that the listener will understand what we are really getting at. Imagine you get stuck in a traffic jam on the way home. "Terrific," you say to your companion, "that means we won't have to sit through the game on TV tonight." If you are a sports fan, this is irony. It is used when there is a discrepancy between what we want or expect, and reality, and occurs widely in such everyday expressions as *clear as mud*, or *Oh great!* when you learn that they'll need your car for yet another day before it's fixed. Election time in my country seems to bring out an epidemic of the expression *Yeah, right.*[17]

Irony is also a well-known literary device. At the beginning of *Pride and Prejudice* Jane Austen was being ironic when she wrote "It is a truth universally acknowledged, that a single man in possession of a good fortune, must be in want of a wife." She really means the opposite—it's women or their mothers who desperately seek a rich single man for marriage. Jonathan Swift was being ironic when he put forward *A Modest Proposal* to solve the problems of starvation and overpopulation in Ireland by eating babies. In *The Ransom of Red Chief*, the short-story writer O. Henry refers to a city: "As flat as a flannel-cake, and called Summit, of course." A crude form of irony is sarcasm, signaled by a sneering tone of voice. Dostoyevsky called it "the last refuge of modest and chaste-souled people when the privacy of their soul is coarsely and intrusively invaded."[18]

Irony depends on theory of mind, the secure knowledge that the listener understands one's true intent. It is perhaps most commonly used among friends, who share common attitudes and threads of thought; indeed it has been estimated that irony is used in some 8 percent of conversational exchanges between friends.[19] Irony can be dangerous if one moves outside one's circle of friends and acquaintances—as a New Zealander I occasionally find myself misunderstood, roughly as a function of cultural distance. I have no doubt that people in Italy or Japan have their own acute sense of irony, but I have the sense that I may have left in those countries a trail of perceived outrageousness.

Irony seems to pose special difficulties for people with autism. Szilvia Papp tells of an otherwise bright 16-year-old boy diagnosed with autistic spectrum disorder who undertook five GCSE exams in Britain, and became extremely upset when his father jokingly said, "If you don't pass you'll have to do them again."[20] Such difficulties extend to other kinds of nonliteral statements, such as the use of metaphor. Papp observes that if told it is raining cats and dogs, the autistic boy will look to the sky to see where these unlikely animals are coming from. Francesca Happé notes that a discussion with a bright autistic child can reveal just how much we use metaphor in everyday speech.[21] For example, asking the child to "give you a hand" meets with the serious reply that she needs both hands and cannot cut one off. Asking her to "stick her coat down over there" is met with a request for glue. Or telling an autistic boy that his sister is "crying her eyes out" leads to anxious peering at the floor to find out where her eyes have gone. Because of the ubiquity of metaphor in everyday language, autistic individuals are unable to follow soap operas, and prefer learning lists of train times to reading fiction.

Theory of mind also allows normal individuals to use language in a loose way that tends not to be understood by those with autism. Most of us, if asked the question "Would you mind telling me the time?" would probably answer with the time, but an autistic individual would be more inclined to give a literal answer, which might be something like "No, I don't mind." Or if you ask someone whether she can reach a certain book, you might expect her to reach for the book and hand it to you, but an autistic person might simply respond yes or no. This reminds me that I once made the mistake of asking a philosopher, "Is it raining or snowing outside?"—wanting to know whether I should grab an umbrella or a warm coat. He said, "Yes." Theory of mind allows us to use language flexibly and loosely precisely because we share unspoken thoughts, which serve to clarify or amplify the actual spoken message.[22]

The language deficits in autism apply primarily to pragmatics—the adaptation of language to social or real-world contexts. In other respects, autistic language may be relatively normal, especially in individuals with the high-functioning form of autism known as Asperger's syndrome. The books that Temple Grandin has written are grammatically correct and reveal a vocabulary that is if anything superior to normal. Her deficits in the use of language are primarily social, although she has learned to compensate by means of dogged attention to how people behave. There may even be respects in which language in people with Asperger's syndrome is superior to normal—in the use of technical language, perhaps. One study shows high-functioning autistic boys to be superior to normal in the naming of pictures of objects.[23] But it is the social function of language, its role in storytelling, gossip, and group bonding that was probably critical to the evolution of language in the first place. Individuals with autism appear to use language primarily to acquire information rather than to share it.

The understanding that true language involves theory of mind, and the sharing of information, now allows us to take a closer look at gestural communication in chimpanzees, and see what extra is needed to turn this into language.

Gestural Origins

I have argued in chapter 4 that the origins of language lie in manual gestures, and the most language-like behavior in nonhuman species is gestural. But it is still not language. Most authors suggest that the deficit is specific to language, a lack of recursive grammar or of what has been rather clumsily called the *faculty of language in the narrow sense* (FLN).[24] In chapter 7 I suggested that this might in turn derive from the inability of nonhuman species, including apes, to travel mentally in time, and generate sequences of past or imagined future events. Here I consider the further possibility that apes may also be incapable of the recursive theory of mind that underlies human conversational language. To examine this, we need to look more closely at how apes communicate using gestures.

According to Michael Tomasello, there are two kinds of ape gesture.[25] One is designed to get another individual to do something. These gestures are small rituals, such as lightly hitting another animal to initiate play, or touching another beneath the mouth to request food, or touching the back of another to initiate piggyback riding. The other kind of gesture is designed to attract attention. A chimp may point to an item of food that's just out of reach. By attracting another individual's attention to it, the chimp may hope that the individual will pick up the food and hand it across. Other attention-getting actions include slapping the ground, throwing things, or poking another chimp.

These behaviors suggest that chimps are at least somewhat sensitive to what's going on in the minds of others, or to what others may do. A chimp may point to an object out of reach to get a person to fetch it for her, but will only do this if she can see that the human is paying attention, which does suggest some awareness of the human's attentional state. Tomasello asserts that chimpanzees will only point for humans, and the failure to observe pointing among chimps in the wild has led to the belief that chimpanzees don't point at all. He suggests, though, that they don't see the point of pointing to each other since they know it won't work. They have learned that humans are cooperative, at least in research settings, and that pointing brings its rewards.[26] Other kinds of gestures between chimps may work, especially if both have something to gain, as in mutual play. In making gestures, then, chimpanzees do seem to be aware of the attentional states of others, whether human or ape, and also aware of the intentions of others. And their gestures are used flexibly and intentionally, rather than as responses induced by events. These qualities are prerequisites for language. Tomasello even refers to them as "the original font from which the richness and complexities of human communication and language have flowed."[27]

But they are not sufficient. Michael Tomasello suggests that the missing ingredient is that of sharing. Chimpanzee gestures are essentially imperative, designed to bring reward or advantage to the gesturer. That is, the chimp is requesting something rather than making a statement. Studies of the use of signs by chimpanzees[28] and bonobos[29] in their interactions with humans have shown that 96–98 percent of their signs are imperative, with the remaining 2–4 percent serving no apparent function—except perhaps one of greeting, or scratching an itch. In marked contrast, human language includes declarative statements as well as imperative ones. We talk in order to share information, rather than merely request something for ourselves.[30]

The declarative function may be evident even in one-year-old human infants, who sometimes point to objects that an adult is already looking at, indicating the understanding that attention to the object is shared. Tomasello gives a number of other examples where the intention is to share rather than to receive gratification. A 13-month-old child watches as her father arranges the Christmas tree. Her grandfather comes into the room, and the child points to the tree for him, as if to say, "Look at the tree, isn't it great?"[31] At 13.5 months, while her mother is looking for a missing refrigerator magnet, a child points to a basket of fruit, under which the magnet is hidden. Such gestures form the basis of language in that they are designed to share information. They also demonstrate that, in development as in evolution, language originates in manual gestures.

This morning I am sitting on a bus. A woman in the seat in front of me is holding a wriggling 18-month-old boy. She tells me that this is his first bus ride, and he is excited. He catches my eye and begins to point to things, including cars and houses, the bus driver, the cord to signal that people want to get off at the next stop. As he points he looks at me. He does not want these things; he wants to share his delight in them with me. I don't think chimpanzees point in this way. I think that children only point like this when there's someone to share the information with.[32]

The toddler has not yet developed full theory of mind, which develops progressively until the age of about four. Pointing to share information nevertheless seems to be an early stage in the emergence

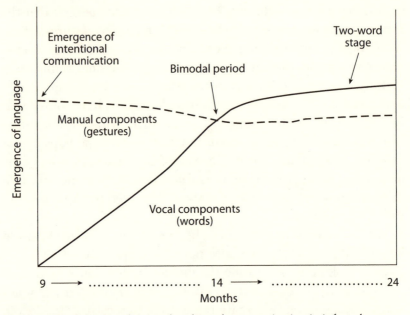

Figure 11. Frequencies of gestural and vocal communication in infants between the ages of 9 and 24 months, based on the work of Volterra et al. (2005). In ontogeny, as I have suggested in phylogeny, intentional gesturers appear well before intentional vocalizations. (Reprinted with permission of Taylor & Francis Group. Figure kindly supplied by Virginia Volterra.)

of both language and theory of mind.[33] It is remarkable that human children already differ from chimpanzees in that this kind of communication seems to be present in one-year-old human infants, but there is no evidence for it in great apes at all. This work provides further evidence that language is built on mental capacity, rather than mental capacity being dependent on language. The mental prerequisites for language, moreover, begin to emerge well before language itself develops. But the emerging theory of mind may be the spur that allows language to unfold.

Chimpanzees, along with bonobos, are our closest living nonhuman relatives, and provide the best estimate of what communication was like prior to the emergence of true language. On the basis of current evidence, then, it seems likely that the additional step that allowed humans to share their thoughts arose in the course of hominin evolution itself, after the split from the apes. Again, the critical ingredient may be recursion, the glue that seems to unite theory of mind, mental time travel, and language itself.

In the next chapter, I place recursion in the context of the debate about whether there is a profound discontinuity between humans and other animals, or whether the difference is one of degree rather than kind.

PART 4

Human Evolution

Whether or not recursion holds the key to the human mind, the question remains how we came to be the way we are—at once so dominant over the other apes in terms of behavior and yet so similar in genetic terms. In chapter 10, I set the problem in terms of the classic debate between Cartesian discontinuity and Darwinian continuity, and then consider some of the steps that made us the way we are. In modern-day science, it is difficult to avoid the conclusion that the human mind evolved through natural selection, although as we have seen, some recent authors, including Chomsky, still appeal to events that smack of the miraculous—a mutation, perhaps, that suddenly created the capacity for grammatical language. This is the "big bang" theory of language evolution, referred to in chapter 4. But of course we do have to deal with the seemingly vast psychological distance between ourselves and our closest relatives, the chimpanzees and bonobos.

In chapter 11 I consider the possible steps by which we became human. The most critical period was probably the Pleistocene, dating from some 2.6 million years ago until around 12,000 years ago. Evolutionary psychologists have supposed that this was the epoch during which the human mind took shape, largely as a consequence of the shift from an arboreal existence to a more terrestrial existence as hunter-gatherers. But as I explain, we were lucky to make it through, since of the 20 or so hominin species so far identified through fossil remains, only one remains.

That species is *Homo sapiens*, who emerged in Africa late in the Pleistocene, some time within the past 200,000 years. That species seems to have been endowed with sufficient qualities to ensure survival, and this is the topic of chapter 12. Some have argued that it was language that made all the difference, as though we talked the Neandertals out of existence, but this view is not in keeping with the Darwinian assumptions of this book. In chapter 12 I suggest that it was not language itself that made the difference. Now read on.

10

The Recurring Question

What a piece of work is a man! how noble in reason! how infinite in
faculty! in form and moving how express and admirable! in action how
like an angel! in apprehension how like a god! the beauty of the world!
the paragon of animals!
　—Shakespeare's *Hamlet* II.2

So spoke Hamlet. Admiration of our own species is certainly one of
our characteristics, although not all authors have been so fulsome.
Blaise Pascal, the seventeenth-century French mathematician, had
a more jaundiced view:

> What a chimera, then, is man! what a novelty, what a monster, what
> a chaos, what a subject of contradiction, what a prodigy! A judge of
> all things, feeble worm of the earth, depositary of the truth, cloaca of
> uncertainty and error, the glory and the shame of the universe![1]

Glory or shame, we cannot but marvel at our human achievements,
although they may ultimately strangle us into extinction on our
hitherto benevolent planet. The variety and ingenuity of human
invention seemingly knows no bounds, ranging from jelly beans
to jumbo jets, hamburgers to *Hamlet*, spears to space probes, internal combustion to the Internet, Beethoven to the Beatles—and
cell phones.

　All of this is in marked contrast to the exploits of our nearest
relatives, the African apes, who are restricted to ever more confined regions of Africa, in conditions that by human standards
are starkly primitive. If they survive at all, it will probably be due
only to the benevolence of humans, and that certainly cannot be

guaranteed. The forests of West Africa are the last stronghold of the African apes in the wild, but mechanized logging means that the numbers of apes (gorillas and chimpanzees) fell by more than half between 1983 and 2002, and there is no end in sight—no end to logging, that is, since the end of the line for the wild chimpanzee seems all too clearly in view. To compound the problem, ebola haemorrhagic fever continues to run unchecked through the chimpanzee population, and hunting for so-called bushmeat has changed from a subsistence activity to a commercial enterprise with the rise of forestry.[2] And yet we share a common ancestry with chimpanzees that goes back only about six or seven million years—an eyeblink on the evolutionary time-scale—and their genetic makeup is something over 98 percent identical to our own.

It is no wonder then that we humans have been tempted to bestow on ourselves some extra quality, perhaps of a nonmaterial sort, that allows us to rise above our ape cousins, closer to angels than to apes. Here is how the Eighth Psalm put it:

> What is man, that thou art mindful of him . . .? For thou hast made him a little lower than the angels, and has crowned him with glory and honour. Thou hast made him to have dominion over the works of thy hands; Thou hast put all things under his feet; All sheep and oxen, yea, and the beasts of the field; The fowl of the air, and the fish of the sea.

Yea indeed. By hovering between ape and angel, we can then indulge in self-glorification for our elevation above the animals, or in self-flagellation for our inability to achieve sainthood, an uneasy equilibrium exploited by religious authorities. In *Paradise Lost*, John Milton wrote:

> The mind is its own place, and in itself
> Can make a Heav'n of Hell, and a Hell of Heav'n.

But even if the path to heaven is an arduous one, are we really justified in supposing that our minds have somehow managed to transcend the physical laws of the universe?

Descartes's Legacy

The idea that we might be possessed of a spiritual transcendence was given scientific and religious respectability by Réné Descartes, sometimes considered the founder of modern philosophy. He was intrigued by mechanical toys, popular at the time, and this led him to wonder whether animals might be reduced to mere machines, whose behaviors could be explained by mechanical principles. He argued that this was indeed true of animals, and for the most part of humans as well, at least with respect to bodily functions and reflexive behavior.

But we humans were unique, he thought, in possessing a flexibility of thought and action that could not be reduced to mechanical principles. This was most evident in language, whose unboundedness defied any attempt to reduce it to deterministic laws, but more generally in free will. We seem to be able to choose what action to take independently of the forces around us. Descartes argued that these freedoms must have arisen from some nonphysical influence that entered the brain through the pineal gland, an organ conveniently located near the middle of the brain so that the incoming signals could be most effectively distributed. That influence could be attributed to God. Because it allowed for free will, it also gave us the opportunity to sin.

Descartes's happy compromise established what is known as *mind-body dualism*, in which the nonmaterial influence that we call mind is separate from the mechanical influences that govern the body, and indeed much of the brain. This was a smart move by Descartes, since it allowed for both the continuation of religion and the development of neuroscience. Yet there is some doubt about whether Descartes was entirely objective in his views, or whether he was simply anxious not to offend the church. Certainly, there were those who disagreed with him. Perhaps the most notable of these was Princess Elizabeth of Palatine, granddaughter of James I of England and niece of Charles I, who challenged the idea that the human mind did not operate according to mechanical laws.

Elizabeth, a devout but tolerant Calvinist, maintained a friendly correspondence with Descartes, but would not allow her letters to him to be published at the time. More than 200 years later her letters were found, and they were eventually published in 1879.

Whatever Elizabeth thought, I suspect that today most people, even in nontraditional societies with no knowledge of Western philosophical thought, would agree with Descartes. Somehow, we don't feel as though we are mere machines. We may well believe that animals are, and this belief may make us feel more comfortable about slaughtering animals for food, or exploiting them for labor or amusement—although we may draw the line at some species, such as family pets, and we may see in the soulful eyes of the chimpanzee a close kindred spirit. In the *Spectator* onetime editor Frank Johnson complained of "boffins" who would reduce humans to robots, proudly declaring his belief in an immortal soul. "Human beings," he wrote, "will always top the earthly hierarchy."[3]

Johnson probably echoes a widespread reluctance to believe that the mind can be reduced to a machine, despite the extraordinary progress of neuroscience and ubiquitous pizza-like images of activity in the brain corresponding to our thoughts and emotions. In the last week of February 2004, the *Reader's Digest* published a survey which revealed that eight out of 10 Australians believe that some people possess psychic powers, and seven out of 10 believe in the afterlife. A majority of people also believe that it is possible to communicate with the dead, and that extraterrestrials have visited the planet. In that same week, Americans voted Australia to be the country with the best image on the planet.[4]

But we shouldn't single out the Australians, since similar statistics could no doubt be compiled from most other societies. One survey shows that about 90 percent of people in the United States believe in God, about 70 percent believe in heaven and the afterlife, and about 58 percent believe in hell.[5] Indeed, a prominent cognitive scientist at Yale University, Paul Bloom, has argued in his recent book *Descartes' Baby* that dualism itself is innate.[6] We have, in other words, a natural-born disposition to believe that mind and body are distinct. Of course, this is not to say that they *are* distinct, but simply that we have been born with the instinct to believe them so. If dualism is wired into our brains, no wonder we have such difficulty accepting a mechanistic view of ourselves. It is an ironic thought, though, that the mechanical functioning of the brain might be what causes it to believe that it is not mechanical.

We should also not judge religion too harshly, since there are sound reasons to suppose that religious belief may itself have been a product of natural selection—not directly, perhaps, but as a consequence of selection for the survival of groups. We humans are fundamentally social creatures, and religion provided one mechanisms for ensuring group cohesion. Religion does pose problems for the theory of evolution, as we shall see below, and the ultimate irony may be that the explanation for religion lies in evolution itself.

Darwin's Heresy

Although Descartes' dualism had had its critics, the most serious challenge came from Darwin's theory of evolution by natural selection, which recognizes no fundamental difference between humans and other species. Not surprisingly, there was antagonism from religious and educational authorities, which continues to this day, a century and a half after the publication of *On the Origin of Species by Means of Natural Selection*. Not for nothing has evolution been dubbed "Darwin's dangerous idea."[7] Darwin himself delayed publication of his book because he knew it would cause trouble. He had been warned, in fact, by the opprobrium that greeted an earlier anonymous publication entitled *Vestiges of the Natural History of Creation*, published in 1844, which dared to suggest that humans might have evolved from primates without divine intervention. Darwin was eventually persuaded to publish only because Alfred Russel Wallace had independently reached similar conclusions, and might publish before he did. Although Darwin had little to say on human evolution until his 1871 book, *The Descent of Man*, the implications were clear. As he correctly surmised, we are descended from the African apes, and we now know with reasonable certainty that we share a common ancestor with the chimpanzee dating from around six or seven million years ago.

Darwin's theory of natural selection is one of the truly major insights in the history of science. The eminent biologist Theodosius Dobzhansky famously wrote that "nothing in biology makes sense except in the light of evolution,"[8] and Darwinian theory is almost universally accepted among biologists, although they may quibble about some of the details. Nevertheless it continues to cause controversy and opposition, especially in parts of the United States. Early in 2004 it was noted that the proposed curriculum in science and math drafted by the Georgia State Board of Education does not include the word "evolution," to the consternation of scientists.[9] There has been persistent pressure, especially in the United States, to introduce an alternative to Darwinian evolutionary theory known as "intelligent design," which is thinly disguised religious doctrine dressed up to look like science.[10]

Intelligent design is explained in an increasingly influential book called *Of Pandas and People*, by Percival Davis and Dean Kenyon. As of August 2005 this book, then in its fifth edition, had sold more than 20,000 copies.[11] Davis and Kenyon write: "Intelligent design means that various forms of life began abruptly through an intelligent agency, with their distinctive features already intact—fish with fins and scales, birds with feathers, beaks and wings, etc."[12] A classic case is the eye, said to be too complex to have evolved incrementally through natural selection. Actually, the eye is not an outstanding example of design, since the retina is installed back to front, and light has to filter through a network of nerve fibers before it reaches the light-sensitive rods and cones, and as a consequence there is a "blind spot" where these fibers gather to leave the eyeball in the optic nerve. It makes you itch to take the thing apart and redesign it.[13]

But it is of course human evolution with which advocates of intelligent design are most concerned. We are so advanced relative to other animals, the argument goes, that we could not possibly have made the journey millimeter by millimeter, through the selection of incremental changes that proved adaptive. Darwinian theory is caricatured as implying that the human condition is the result of random variation, as plausible as the possibility that an ape, given a typewriter, might adventitiously produce the plays of Shakespeare. Of course the theory of natural selection does depend on random variation, but selection of those variations that lead to increased biological fitness leads to systematic progression toward more adaptive forms. The trick is that evolution is a cumulative process, leading incrementally and inexorably toward greater fitness. One end product was a Shakespeare who did write all those plays.[14] The real problem with intelligent design, though, is that it needs an intelligent designer, and its advocates provide no information on where that designer is to be found, or how to approach her to fix up that little problem with the eye. Merely to postulate an all-knowing, all-encompassing intelligent designer is bad science, since it does not explain anything, and there seems no way to refute it.

Not all religions accept intelligent design. Writing in the 18 January 2005 issue of *L'Osservatore Romano*, the official Vatican paper, Fiorenzo Facchini argued that intelligent design belongs to the realms of philosophy and religion, but not of science. He stated that "it is not correct from a methodological point of view to stray away from the field of science while pretending to do science." The Vatican has for many years tolerated the teaching of evolutionary theories, and in 1950 a papal encyclical officially permitted Catholics to discuss Darwin's theory of evolution. In another development, on 20 December 2005 federal district court judge John Jones III ordered the schools in Dover, Pennsylvania, to remove references to intelligent design from the science curriculum, on the grounds that it is not science.

Part of the reason for resisting Darwinian theory is that it seems to dislodge humans from the pedestal of superiority. We should not be too complacent, though. Human society is witness to extraordinary accomplishments, but most individual humans have little understanding of the miracles that surround them. Most of us would be utterly helpless if transported back to the African savanna of a million years ago. Without matches, we would probably have difficulty lighting a fire to keep predators at bay, let alone constructing a helicopter to get us out of there. There are probably only a handful of people who really understand the general theory of relativity, or the recent proof of Fermat's last theorem. Alfred Russel Wallace, Darwin's rival, was so taken with the difference between educated Europeans and primitive "savages" that he was prompted to invoke divine intervention to explain the difference—he wrote that "natural evolution could only have endowed the savage with a brain a little superior to that of an ape."[15] Darwin himself was appalled, and wrote to Wallace, "I hope you have not murdered too completely your own child and mine."[16]

In contrast to the theory of intelligent design, the theory of evolution by natural selection is genuine science in that it is open to refutation, as Darwin himself realized. In the sixth edition of *The Origin of Species*, published in 1872, he wrote: "If it could be demonstrated that any complex organ existed, which could not possibly have been formed by numerous, successive, slight modifications, my theory would absolutely break down. But I can find no such case." Descartes had suggested the pineal gland as the organ that might account for human uniqueness, but this turned out to be without foundation, although some parapsychologists still believe that it may be responsible for telepathy and other extrasensory powers. The nineteenth-century anatomist Richard Owen maintained that a brain structure known as the hippocampus minor was unique to humans, but Darwin's friend and protagonist Thomas Henry Huxley disproved this by showing that all apes possess this structure. The role of the hippocampus minor was ridiculed, albeit in a befuddled way, by Charles Kingsley in his 1886 book *The Water Babies*:

> You may think that there are other more important differences between you and an ape, such as being able to speak, and make machines, and know right from wrong, and say your prayers, and other little matters of that kind; but that is a child's fancy, my dear. Nothing is to be depended on but the great hippopotamus test.

But what exactly *is* the hippocampus minor? Neuroscientists are familiar with the hippocampus, a structure in the brain that plays an important role in memory, and others may identify the hippocampus with that lovely undulating creature, the sea horse, whose feminist principles have ensured that it is the male who carries the unborn young. In the nineteenth century, the hippocampus minor was identified as a ridge in the floor of the rearward horn of the lateral ventricle, and was clearly distinguished from the hippocampus itself, then called the hippocampus major. Owen's idea that the hippocampus minor might be important probably owes something to the famed second-century anatomist Galen, who thought that the faculties of the mind resided not in the matter of the brain itself, but rather in the ventricles, which are the fluid-filled spaces in the brain. In any event, after the spat between Owen and Huxley, the term "hippocampus minor" disappeared, and was replaced by its original name, "calcar avis," meaning cock's spur. It now lives only in the most obscure corners of anatomy texts, and probably doesn't do much for us at all.[17]

Of course if we do discover some complex organ that exists in humans but is not present in the other apes, this would indeed pose problems for Darwinian theory. Such a discovery seems unlikely. There can no longer be any serious doubt that we share our most recent common ancestry with the chimpanzee and bonobo, and slightly earlier ancestry with other great apes. Going further back we share ancestry, albeit ever diminishing, with monkeys, mammals, and ultimately all living creatures. Indeed molecular analysis tells us that we are biologically closer to the chimpanzee than the chimp is to the gorilla, outward appearances notwithstanding. Any uniqueness we may claim has come about, not through the magical insertion of some new organ, or through divine intervention, but rather through evolutionary tinkering. This no doubt involved the modification of growth patterns, and the handy trick of using organs that evolved for one purpose to achieve quite different ends. Just as the nose was modified to become the elephant's trunk, and the forelimbs to become the wings of birds, so nature has taken the body and brain of an ape and turned it into a human. Just another ape, really, albeit one with some interesting properties.

Many of those properties have to do with what we are pleased to call *mind*. It is sometimes claimed that only humans have minds, or consciousness, but this is surely wrong. Other primates are clearly able to think. So indeed are other mammals, cetaceans, and birds. But it is probably true that we humans have evolved ways of thinking that are unique, although derived from mental structures that were already present in our forebears. Moreover, our minds are not the phantoms of Cartesian dualism. Modern neuroscience is relentless in showing us that what we think of as mind and consciousness are due to the workings of the physical brain. Of course, we do not yet understand how consciousness itself comes about, although there is much speculation, so the identity of mind and brain must still be considered a working hypothesis. From a scientific viewpoint, the only real contender for the seat of the mind, or even the soul, is the brain.

So how do our minds—or brains—differ from those of other species?

Is Recursion the Answer?

In this book, I have argued that recursion might provide the key. Recursion might be said to capture several other properties previous claimed as unique to humans—language, episodic memory, mental time travel, and theory of mind. It also permits a degree of continuity, since each of those properties has precursors identifiable in nonhuman species. Language no doubt grew out of animal communication (*pace* Chomsky), perhaps more directly from gestural than from vocal communication, as I argued in chapter 4. Episodic memory and mental time travel might be considered refinements of memory capacities already evident in other species. And apes do show at least some glimmerings of a theory of mind.

Some have argued that it was not language per se that gave rise to the distinctive character of the human mind, but rather the ability to think in symbols. This was the theme of Terrence Deacon's (1997) book *The Symbolic Species*, and more recently of an article by Derek C. Penn, Keith J. Holyoak, and Daniel J Povinelli, mentioned in chapter 8. The use of abstract symbols does of course characterize human language, especially speech. I argued in chapter 4, though, that the use of abstract symbols in language was more a matter of expedience than a distinctive component of the human mind. Chimpanzees and bonobos can be taught to use abstract symbols. Use of symbols has of course been exploited in mathematics and engineering, leading to the development of science and complex manufacture. But these are the accomplishments of Western civilization, and are generally foreign to indigenous peoples. In origin, at least, language is perhaps more a question of embodiment than of symbolic manipulation,[18] and if it evolved from manual gestures, and indeed can persist in that form, we should look to processes not involving abstract symbols as the key to the human mind. That is why I suggest recursion as a possibility.

But is recursion truly a magic bullet, turning preexisting processes from cramped stereotypy to the soaring creativity we see in stories, poetry, art, music, and dance, not to mention the more oppressive dominance of machines and skyscrapers? The evidence reviewed in chapters 8 and 9 that great apes may have a limited ability to read the minds of others, whether expressed in acts of deception or in communicative pointing, or even in scrub jays' apparent ability to prepare for a future event. We will need further careful analyses to determine whether such cases truly reflect a recursive ability, or whether they can be explained more simply in terms of association. One possibility, suggested in the previous two chapters, is that recursive processing can be discerned in some animal behaviors, but does not extend beyond a single level of embedding. Humans have the ability to share, to tell stories within stories, to develop devious social strategies, to play chess or poker, even to do mathematics and write computer programs that call recursively on other programs—all of these suggest runaway recursion that goes beyond first-order intentionality to perhaps as high as fifth or sixth order.

All of this need not imply a profound discontinuity, or threaten Darwinian principles. An appropriate biological analogy might be flight. Kangaroos can hop, dolphins can leap out of the water, even humans can create a meter or two of distance between themselves and the earth—although we now do rather better since we have constructed flying machines, inelegant though they are. But the evolution of wings created a profound discontinuity from incremental changes to limbs initially adapted to terrestrial movement. One small hop for animals became a giant leap for birds.[19]

From Descartes to Chomsky, the case for a discontinuity between us flightless humans and other species has been based primarily on the supposed uniqueness of language. Although Chomsky himself has made no claims about the human soul, he is in other respects a self-confessed Cartesian,[20] and has long argued that language is uniquely human, primarily because of its recursive properties. We saw in chapter 2, though, that some languages, such as that of the Pirahã, may not make use of recursion. The prior significance of recursion may therefore lie, not in language itself, but rather in the nature of human thought that guides language and supplies much of its content. I argued in chapter 5 that the act of mentally reliving past events is a recursive process, analogous to calling a subroutine into a main routine. We can do this beyond the level of first-order recursion when we imagine imagining yesterday what we had planned to do today. In chapter 6 I extended this notion to the imagining of future events, or the construction of stories, and in chapter 7 I argued that these properties underlie at least some of the characteristics of language. The recursive tangle of human relationships also supplies the material for gossip, one of our favorite pastimes. The recursive structure of at least some languages, then, owes something to the recursive manner in which we construct episodic scenarios.

In chapter 8 I discussed yet another recursive function critical to the human condition. The capacity to know what others are thinking is recursive in that the mental processes of others are called into our own thoughts, and guide our social interactions. Rudimentary theory of mind can perhaps be inferred from the behavior of other species, but in humans goes beyond first-order recursion. The knowledge that we not only know what others are thinking, but also know that they know what we are thinking, may underlie what is perhaps the most fundamental human trait, the capacity to share. In chapter 9 I pointed out that language is just one manifestation of this. Language provides for the sharing of knowledge.

In the next two chapters, I try to set these recursive functions into human evolution itself.

11

Becoming Human

Human life is a sad show, undoubtedly: ugly, heavy and complex.
—Gustave Flaubert

We are, let's face it, great apes, sharing most recent common ancestry with the chimpanzee and bonobo. The other great apes are the gorilla and orangutan. We set ourselves on the track to humanity some six or seven million years ago, when the so-called hominins[1] split from the line leading to modern chimpanzees and bonobos. It was a not altogether successful venture, since nearly 20 hominin species have been identified from fossil remains, but only one species of hominin remains on the planet—see figure 12. That species is *Homo sapiens*. Lucky you.

In this book, I have argued that the mind of *Homo sapiens* is possessed of a recursive property unique among extant creatures, providing the creative potential for such diverse activities as reconstructing past episodes or imagining future ones, telling stories, creating music or art, and manufacturing edifices and complicated machines. The other great apes may possess some degree of flexibility in communication, especially through manual and bodily gesture, but their communications have virtually none of the generativity of human language. They do not tell stories. Whether they have episodic memory or can plan episodic futures remains a matter of contention. Nevertheless there is no indication that apes truly have a sense of their past mental lives, or that they can construct episodic futures, such as a meeting with a lover after a quarrel and imagining how it will play out.

Can great apes understand what is going on in the minds of others? They can surely gauge the emotional state of another individual,

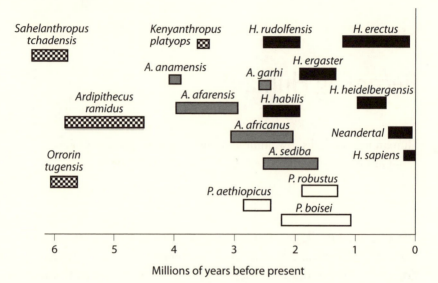

Figure 12. Species identified arising since the split from the chimpanzee line, and identified as bipedal hominins. *A = Australopithecus, P = Paranthropus, H = Homo.*

and perhaps take the visual perspective of another individual. Some evidence, reviewed in chapter 8, suggests that a chimp may have some understanding of what another individual knows or believes, although this remains controversial. There is no evidence that an ape's understanding extends to higher-order recursion, such as the understanding that a watching individual understands that the watched individual understands this understanding! Of course these assertions are open to challenge.

We can assume, then, that the capacities of present-day great apes do not underestimate the capacities of the common ancestors of ourselves and the other apes. If anything they overestimate them, since chimpanzees have also evolved over the past six or seven million years, and I don't imagine that they once had but have since lost the powers of language, memory, or recursive theory of mind that we humans have evolved. It follows that these capacities must have emerged over that period. In this and the next chapter, then, I trace the stages of hominin evolution that may have led to our distinctive mental capacities.

Let's start with what some regard as the defining characteristic of the hominins—bipedalism.

Standing Up for Ourselves

> When the first humans rose to their hind feet, wobbled for millennia, finally stood upright, achieved the erect posture, it must have been a sorry day but a touch of grandeur was on it.
> —Gustav Eckstein, *The Body Has a Head*

To reverse the slogan of the pigs on George Orwell's *Animal Farm*, "Two legs good, four legs bad," we pride ourselves that we differ from the other apes—and from pigs—in that we strut about on two legs.

The early hominins appear to have been at least facultative bipeds, which means that bipedalism was an optional form of locomotion, shared with the ability to climb trees. Facultative bipedalism is contrasted with the obligate bipedalism of modern humans, where there is no alternative way of getting about.[2] The earliest fossil tentatively identified as a hominin, known as *Sahelanthropus tchadensis*, was discovered in Chad in Central Africa, and is dated at between six and seven million years ago.[3] This was very close to the time of the chimpanzee-hominin split, estimated at between 6.3 and 7.7 million years ago by a technique known as DNA hybridization.[4] Early indications from the mounting of the skull on the backbone are that this creature was capable of bipedal walking. Another early fossil, *Orrorin tugenensis*, dating from between 5.2 and 5.8 years ago,[5] is perhaps more securely identified as a facultative biped, as is *Ardipithecus ramidus*, dating from between 5.4 and 4.4 million years ago. Later hominins, including those known as australopithecines, were also bidedal walkers, as evident from the famous footprints attributable to *Australopithecus afarensis*, popularly known as Lucy, and dated at around 3.5 million years ago.[6]

Chimpanzees, bonobos, and the slightly more distant gorillas are so-called knuckle-walkers, which means they are basically quadrupeds, using the forearms as legs, with the knuckles rather than the

palms of the hand touching the ground. Because chimps and bonobos are our closest living relatives, it has generally been assumed
that the ancestor we share with them, and indeed the earlier ancestor we share with the gorilla, must also have been knuckle-walkers.
That is, the earliest hominins progressed from knuckle-walking to
bipedalism.

This idea has recently been challenged. To understand how the
different postures of the great apes came about, we have to go back
to the trees, whence we came. The most arboreal of the great apes is
the orangutan, which is more distantly related to us than either the
chimpanzee or gorilla. Nevertheless its body morphology is closer
to that of the human than is that of chimpanzee or gorilla. In the
forest canopy of Indonesia and Malaysia, orangutans typically
adopt a posture known as *hand-assisted bipedalism*, supporting
themselves upright on horizontal branches by holding on to other
branches, usually above their heads. They stand and clamber along
the branches with the legs extended, whereas chimpanzees and gorillas stand and move with flexed legs. Chimpanzees and gorillas
may have adapted to climbing more vertically angled branches, involving flexed knees and a more crouched posture, leading eventually to knuckle-walking as the forested environment gave way to
more open terrain. If this scenario is correct, our bipedal stance may
derive from hand-assisted bipedalism, going back some 20 million
years. Knuckle-walking, not bipedalism, was the true innovation.[7]

It is *Ardipithecus ramidus*, though, who gives the clearest picture of what the common human-chimpanzee ancestor must have
been like—see figure 13. Ardi, as she is known, has provided the
most complete skeleton of the early hominins. Although her pelvic
structure suggests she was a facultative biped, her foot was still
adapted to grasping, with an opposable big toe. Her hand was
closer to that of humans, and indeed of earlier primates, than to
that of the chimpanzee or gorilla, which were adapted to knuckle-
walking. Like the human hand, Ardi's hand could bend backward
at the wrist, so she could move on flat hands and feet when on horizontal branches, whereas the chimp and gorilla have stiff wrists
to support knuckle-walking.[8] As Owen Lovejoy suggests, Ardi,
or rather her species, "has been bipedal for a very long time."[9]

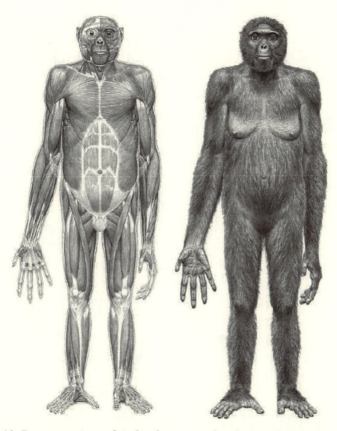

Figure 13. Reconstructions of *Ardipithecus ramidus* (from *Science*, 2009, 326, p. 36, reprinted with permission of the illustrator, Jay Matternes.).

This is further evidence that the chimpanzee and gorilla, with their knuckle-walking, were something of a sideshow to the progression of bipedalism.[10] It seems we did not evolve from knuckle-walkers, but represent the end point of a gradual transition from tree-climbers to bipedal walkers.

Bipedalism became obligate rather than facultative from around two million years ago, and as we shall see, it was from this point that the march to humanity probably began. That is, our forebears finally gained the capacity to walk freely, and perhaps run, in open terrain, losing much of the adaptation to climb trees and move about in the forest canopy. Even so, it remains unclear just why bipedalism was retained. As a means of locomotion on open terrain, it offers no obvious advantages. Even the knuckle-walking chimpanzee can reach speeds of up to 48 km per hour,[11] whereas a top athlete can only run at about 30 km per hour. Other quadrupedal animals, such as horses, dogs, hyenas, or lions, can easily outstrip us, if not leave us for dead. One might even wonder, perhaps, why we didn't hop rather than stride, emulating the bipedal kangaroo, which can also outstrip us humans.

Bipedalism is in many ways disadvantageous. It gives rise to back and neck problems, hemorrhoids, hernias, and the excessive pain of giving birth.[12] Sciatica can be blamed on the fact that the spinal nerves and spinal discs are too close together, again a result of adaptations for our two-legged stance. Human babies take a long time to learn to walk, which makes them especially vulnerable to the sorts of predators that like to eat babies. A tragic example is the infamous dingo that carried off Lindy Chamberlain's 10-week-old baby daughter, Azariah, at Uluru[13] in Central Australia. Having just two legs for walking means that we are especially helpless if we lose the use of one of them—hopping is highly inefficient compared with the ease with which a three-legged dog, say, can still get around.[14] So much for intelligent design! The impression is that the two-legged model, like the Ford Edsel, was launched before the market was ready for it. Or before it was ready for the market, perhaps.

But there must have been adaptive advantages. These may have had to do in large part with the freeing of the hands and arms for manipulative purposes, and perhaps leading eventually to recursive manufacture. As we shall see, though, manufacture emerged very late in hominin evolution, and cannot account for the pressures toward bipedalism, which may go back to well before our common ancestry with the great apes, if the above scenario is correct. The freeing of the hands would have been adaptive for other purposes, though, such as grooming, carrying things, or fighting.

Throwing

One possibility is that standing up allowed us to throw things, and so develop superior defensive and hunting skills. Charles Darwin suggested this when he wrote: "In throwing a stone or spear a man must stand firmly on his feet."[15] In our modern sedentary lives we may have lost some of this ability, although you only need to watch professional baseball players or cricketers, or American-style footballers, to see that some of us, at least, are still capable of prodigious feats of throwing. The ability to throw, with potentially lethal accuracy, may be more widespread among hunter-gatherer societies than among us sedentary city-dwellers. The eighteenth-century explorer J. W. Vogel wrote that the Hottentots of southwestern African "know how to throw very accurately with stones. . . . It is also not rare for them to hit a target the size of a coin with a stone at 100 paces."[16] Australian aboriginals were also said to be able to throw stones with enough accuracy and force to bring down wallabies and flying birds, dislodge nuts from the baobab tree, and knock fledgling birds out of high nests.[17]

In his book *The Throwing Madonna*, William H. Calvin suggested that it was women who were the throwing experts.[18] They stood as proud Amazons, holding their infants in their left arms so that they were snuggled close to the beating heart, and used their right arms to hurl objects at dangerous predators. This would have favored the selection, not only of a bipedal stance, but also of circuits in the brain that could program the precise timing needed for accurate throwing. Since the right arm is largely controlled by the left brain, this may have set the stage for the later evolution of speech, which also requires precise timing, and which is programmed in the left brain in the great majority of people. One might quibble, I suppose, that men seem to be better than women at throwing, or at least more likely to indulge in pursuits that involve accurate throwing, although this impression can be partially dispelled by watching an elite women's cricket team.

Eduard Kirschmann, in his book *Das Zeitalter der Werfer* (*The Age of Throwers*),[19] restores male vanity by arguing that accurate throwing was indeed in the hands of the men, in part perhaps to protect the mothers holding their infants, and in part also to hunt prey. One imagines that selection of the best throwers was not solely a matter of survival against threat, but that sexual selection may also have played a role. Witness the spectacle of young men parading their skills in sporting arenas, be it in baseball, cricket, rugby football, discus, javelin, or an activity known as Australian

Rules.[20] Among the early hominins, the upright stance would also have freed the hands for carrying the objects to be thrown, in case one encountered some predatory beast in one's path.

Certainly, there are good reasons to suppose that the upright stance would have greatly increased the leverage needed for powerful throwing. Watch a baseball pitcher and you see that the throwing motion begins in the legs and feet, and unfolds through hips, torso, shoulder, arm, elbow, wrists, and fingers.[21] This whole-body action maximizes kinetic energy. Of course one might still argue that adaptations for throwing don't really explain bipedal *walking*,[22] but Kirschmann also points out that the flexibility of the wrist required for powerful throwing would have hindered the use of the hands for locomotion. Mary Marzke makes the point that if bipedalism was an adaptation for locomotion, we should have evolved legs that were better designed for that purpose—like those of an ostrich, perhaps. In contrast, our legs are much sturdier, and the knee includes a locking device that has little to do with locomotion itself. These features may well have evolved to provide a more stable launching platform, not for locomotion per se, but for throwing and clubbing.[23]

The idea that we evolved to throw also helps explain the mystery of the seeming perfection of the human hand, described by Jacob Bronowski as "the cutting edge of the mind."[24] Adding to the power and precision of throwing are the sensitivity of the fingers, the long opposable thumb, and the large area of cortex involved in control of the hand. The shape of the hand evolved in ways consistent with holding and hurling rocks of about the size of modern baseballs or cricket balls, missile substitutes in modern pretend war. In real war, hand grenades are about the same size. Our hands have also evolved to provide two kinds of grip, a precision grip and a power grip, and Richard W. Young suggests that these evolved for throwing and clubbing, respectively.[25] Not only do we see young men throwing things about in sporting arenas, but we also see them wielding clubs, as in sports such as baseball, cricket, hockey, or curling. In various forms of racket sports, the skills of clubbing and throwing seem to be combined.

Comparisons between the hand of the chimpanzee and that of

early hominins suggest increasing hominin adaptations for throwing, although as we have seen, the chimpanzee hand may be a misleading comparison, as it probably adapted to knuckle-walking after the split from the common ancestor. Nevertheless the remarkably complete 3.3-million-year-old skeleton of a juvenile *Australopithecus afarensis*, recently found in Dikika, Ethiopia, has curved fingers like those of a chimpanzee, apparently adapted for holding onto branches. The lower body of this little girl is adapted for bipedal locomotion, but her upper body is still in many respects apelike.[26] Whether she threw things is not known.

It wasn't as though the capacity to throw started entirely from scratch—or even from scratching—since present-day primates can also throw things, although not as accurately or powerfully as modern humans can. Capuchin monkeys, found in South and Central America, can throw stones at both moving and stationary objects, and have some semblance of both power and precision grips for throwing. In one study, they proved quite accurate at throwing stones into a bucket containing either peanut butter or a sweet syrup, the reward for accuracy being opportunity to lick the stone afterwards. They nearly always threw over-arm, about half the time from an upright stance. The female monkeys were just as accurate as the males, although both were less proficient than the humans who were tested.[27]

Chimpanzees also hurl objects, such as branches of trees, in self-defense. You may need to be on your best behavior when visiting the zoo—Charles Darwin wrote: "As I have repeatedly seen, a chimpanzee will throw any object at hand at a person who offends him."[28] Watch out too when visiting the Cape of Good Hope, since Darwin also mentioned a baboon there who not only threw missiles at people, but also prepared missiles made of mud for the purpose. That animal has no doubt passed on, but may well have left his legacy among his descendants. When the bonobo Kanzi, our guest in chapter 3, was shown how to make simple flake tools, he did not adopt the hammering style used by the hominins of around two million years ago, but flung the stones at a hard surface, so that the flakes broke off with the impact.[29] These great apes do not throw with the precision and power of a baseball or cricket player, but their ability and inclination to throw suggests that the subsequent emergence of throwing in hominins had a platform upon which to build.

Paul Bingham has argued that one of the characteristics that have reinforced social cohesion in humans is the ability to kill at a distance.[30] Human societies can therefore be rid of dissenters in their midst, or threats from outside, with relatively little threat of harm to the killer! Nevertheless the dissenters, or the rival band, may themselves resort to similar tactics, and so began an arms race that has continued to this day. It started, perhaps, with the throwing of rocks, followed in succession by axes, spears, boomerangs, bows and arrows, guns, rockets, bombs, and nuclear missiles, not to mention insults. Such are the marks of human progress. People are still quick to regress to throwing things to express aggression, as angry crowds in the trouble spots of the world throw stones, rocks, or bottles at those they love to hate. Throwing stones seems even to have infiltrated the literary establishment. George Bernard Shaw expressed his distaste for William Shakespeare as follows:

> With the single exception of Homer, there is no eminent writer, not even Sir Walter Scott, whom I can despise so entirely as I despise Shakespeare when I measure my mind against his. . . . It would positively be a relief for me to dig him up and throw stones at him.[31]

Whether or not it was throwing that sustained bipedalism in an increasingly terrestrial existence, it does at least illustrate that bipedalism frees the hands for intentional and potentially skilled action. It allows us to use our hands for much more than simply chucking stuff about. Moreover, our primate heritage means that our hands and arms are largely under intentional control, creating a new potential for operating on the world, instead of passively adapting to it. Once freed from locomotory duty, our hands and arms are also free to move in four-dimensional space, which makes them ideal signaling systems for creating and sending messages. Shaw might have considered the possibility of gesturing to express his contempt—a two-finger salute might have been just as effective as hurling stones, assuming of course that Shakespeare recognized its significance.

Indeed, one might be tempted to suppose that bipedalism was driven by the emergence of language itself, especially if language evolved from manual gestures, as I maintained in chapter 4. Nevertheless there is little else in the fossil record to suggest any substantial progress toward the human mind for the first four or five million years following the split from the great apes. For all the freedom of the hands, our early ancestors seem to have been very slow to make tools—at one time considered the mark of humanity. They may well have used twigs or stones as makeshift implements, much as modern chimpanzees do, but there is little evidence for the systematic construction of tools until well after the split from the chimpanzee and bonobo. Their bipedalism, moreover, was still relatively clumsy and inefficient. They retained some arboreal characteristics, with long arms and relative short legs, and their brains were little or no larger than those of modern chimpanzees. It was not until the Pleistocene that we begin to see tangible signs of change, and a steady march towards humanity.

The Pleistocene

Toward the end of the Pliocene, which originated some 5.3 million years ago, the earth underwent global cooling, and the ensuing Pleistocene gave rise to a series of crippling ice ages. The Pleistocene was once dated from 1.81 million years ago, but according to a recent vote has been redated to stretch from 2.588 million years ago to some 12,000 years ago.[32] The Pleistocene also marks a shift from a largely wooded environment to more open savanna, forcing further adaptations toward a terrestrial rather than an arboreal existence. Adaptation to the Pleistocene resulted in a new genus, called *Homo*, with characteristics rather different from the earlier australopithecines.

The earliest members of the genus were *Homo habilis* and *Homo rudolfensis*, although it has been suggested that these species had not really attained *Homo*-like characteristics, and should be classified still as australopithecines.[33] Another hominin, named *Australopithecus sediba*, has recently been unearthed in South Africa, dating from around 1.9 million years ago, and may be truly transitional between *Australopithecus* and *Homo*.[34] *Homo ergaster*, who emerged a little over 1.8 million years ago, clearly belongs to the genus, as does the Asian variant *Homo erectus*. Later variants were *Homo antecessor, Homo heidelbergensis, Homo Neandertalensis* (Neandertals),[35] and eventually the intrepid survivor, *Homo sapiens*. The features that identify *Homo* as the escapee from apedom, and the increasing bearer of attributes we are pleased to call human, probably have to do with the very specific conditions and dangers of the Pleistocene.

An especially dangerous feature of the savanna was the presence of large carnivorous animals, whose numbers peaked in the early Pleistocene. They included at least 12 species of saber-tooth cats and nine species of hyena.[36] Our puny forebears had previously been able to seek cover from these dangerous predators in more forested areas, and perhaps by retreating into water, but such means of escape were relatively sparse on the savanna. Not only did the hominins have to avoid being hunted down by these professional killers, with sharp teeth and claws, and immense speed and strength, but they also had to compete with them for food resources.

Consider then the plight of our forebears, increasingly forced to compete on the savanna with killer cats. One may wonder why we did not evolve to compete directly with these dangerous predators on their own terms, evolving greater speed, strength, and killing capability—as recommended by the king in Shakespeare's *King Henry V*:

> Then imitate the action of the tiger;
> Stiffen the sinews, summon up the blood,
> Disguise fair nature with hard-favoured rage:
> Then lend the eye a terrible aspect;
> Let it pry through the portage of the head,
> Like the brass cannon; let the brow o'erwhelm it,

As fearfully as doth a galled rock
O'erhang and jutty his confounded base,
Swilled with the wild and wasteful ocean,
Now set the teeth, and stretch the nostril wide;
Hold hard the breath, and bend up every spirit
To his full height!

But our forebears were scarcely preadapted for direct physical competition with the tiger. Coming from forested environments, the early hominins were not built for excessive strength or aggression.

Nor were they built for speed; otherwise an alternative strategy might have been to evolve more efficient ways to escape from predation, emulating not the action of the tiger, but rather the bounding grace of the antelope. The facultative bipedalism of the earlier australopithecines did give way to obligate bipedalism, with a full striding gait and limbs better adapted to running. But this was scarcely sufficient to allow escape from a hungry lion, and seems to have been more specialized for endurance running, perhaps allowing them to compete with other scavengers in the early Pleistocene, and even run some mammalian prey to exhaustion in the heat.[37] Yet again, our ancestors were not built for airborne flight, and hairy arms and heavy bodies would have required considerable modification before they could take to the skies to escape a mastodon. Flight, when it did emerge in our species, came much later, in part, I'm pleased to say, through the efforts of a New Zealander.[38]

The solution seems to have been to depend on social intelligence and the development of cooperation. Darwin put it like this:

The small strength and speed of man, his want of natural weapons, &c, are more than counterbalanced, firstly by his intellectual powers, through which he has formed for himself weapons, tools, &c, though still remaining in a barbarous state, and, secondly, by his social qualities which lead him to give and receive aid from his fellow-men.[39]

Sarah Blaffer Hrdy has argued that social bonding evolved first in the context of child rearing.[40] She points out that great apes are loathe to allow others to touch their infants during the first few months, whereas human mothers are very trusting in allowing others to carry and nurture their babies. This is evident not only in daycare centers, but in the extended families units that characterize many peoples of the world. Among New Zealand Maori, for instance, initial teaching and socialization is based on a larger unit known as *whanau*, which is the extended family, including children, parents, grandparents, cousins, uncles, aunts, and often beyond. The understanding of whanau is recursive, looping back many generations.

The close contact between infants and other individuals besides the mother would not only have enhanced survival, but encouraged the cooperative spirit that would later enable these forebears of humans to trust others with their own infants. A spiral of mutual trust and caring would therefore extend down the generations. The exposure of infants to a variety of others would also foster theory of mind, teaching infants to gauge the intentions of others, and learn whom to trust. Hrdy argues that these characteristics predated the expansion of the brain, discussed below, and may even have established the initial conditions favoring an increase in brain size, and the evolution of such traits as social learning, teaching, and language itself.

The hominins therefore built on their primate inheritance of intelligence and social structure rather than on physical attributes of strength or speed.[41] This is what might be termed the third way, which was to evolve what has been termed the "cognitive niche,"[42] a mode of living built on social cohesion, cooperation, and efficient planning. It was a question of survival of the smartest.

Nevertheless it is unlikely that human intelligence can be explained simply in terms of the response to ecological challenges. Our Pleistocene forebears seemed to have conquered the threats posed by the inhospitable environment, but then discovered a further threat—themselves. Paraphrasing an earlier seminal paper by Nicholas K. Humphrey,[43] Richard D. Alexander writes that "the real challenge in the human environment throughout history that affected the evolution of the intellect was not climate, weather, food shortages, or parasites—or even predators. Rather, it was the necessity of dealing continually with our fellow humans in social circumstances that became ever more complex and unpredictable as the human line evolved."[44] Humans have proven as adept at killing each other as at killing nonhuman predators—in fact more so, if you consider the extraordinary range of lethal weapons we humans have invented. Nevertheless successful adaptation must have depended as much on cooperation as on competition, leading to what has been called "runaway social selection." Our lives depend on a subtle calculus of sharing and greed—of left-wing socialism and right-wing individualism, if you like.

Richard Wrangham has suggested that the secret of hominin evolution originated in the controlled use of fire, which supplied warmth and protection from hostile predators. From around two million years ago, he thinks, *Homo erectus* also began to cook tubers, greatly increasing their digestibility and nutritional value. Cooked potatoes, I'm sure you will agree, are more palatable than raw ones. Other species may have been handicapped because they lacked the tools to dig for tubers, or the means to cook them. Cooked food is softer, leading to the small mouths, weak jaws, and short digestive system that distinguish *Homo* from earlier hominins and other apes. Cooking also led to division of labor between the sexes, with women gathering tubers and cooking them while the men hunted game. At the same time, these complementary roles encouraged pair bonding, so that the man can be assured of something to eat if his hunting expedition fails to produce meat to go with the vegetables.[45]

Could social bonding have truly evolved around the campfire, with barbecued meat and vegetables, and gestured storytelling? It's an attractive scenario, but it remains speculative in the absence of evidence for the use of fire early in the Pleistocene. The earliest convincing evidence dates from only about 800,000 years ago.[46]

Let's now examine some of the tangible evidence that the march from ape to human really began in the Pleistocene. I begin with the very organ that we think of (and with) as supplying the attributes that underlie our uniquely recursive mode of consciousness.

The Brain

> Man by his large and choice brain has occupied the earth, and the
> house mouse, the field mouse, the house rat, the bed bug, certain
> flukes, worms, mites, ticks, lice, fleas, all have followed man
> around, and may, to that extent, be said to have acknowledged
> that large and choice brain, and to have cast their lot with it.
> —Gustav Eckstein, *Everyday Miracle*

I have already suggested in earlier chapters that functions such
as language, with the vast vocabulary of words and underlying
concepts, and episodic memory, with its storehouse of individual
episodes, placed new demands on neural storage, and may have
driven the dramatic increase in brain size that occurred during the
Pleistocene. Recursion, too, must have added to the pressure, since
it requires hierarchical structure, enhanced short-term memory,
and sequential programming.

But before we congratulate ourselves on our big-headedness, we
do need to inject a note of humility. You might have thought that
we humans have the largest brains of all in the animal kingdom,
given that we consider ourselves to be top of the earthly hierarchy
in terms of intelligence, but in fact we have to defer to the elephant
and the whale, whose brains are more than four times as big as
our own. The human brain is about the same size as that of a dolphin. Fortunately, the absolute size of the brain is not very revealing about intelligence. Large animals need large brains simply to
control those big bodies, and deal with all of the information that
arrives from their large surfaces.[47] A more revealing index, then,
may be the ratio of brain size to body size.

Here we come out rather better than elephants and whales, and
better than our great ape cousins. Our brains weigh about 2.1
percent of our body weights, while those of the chimpanzee and
bonobo are about 0.61 percent and 0.69 percent respectively. The
gorilla weighs in at about 0.64 percent and the orangutan at about
0.55 percent. The figures for the bottlenose dolphin, Asian elephant, and killer whale are 0.94 percent, 0.15 percent, and 0.094
percent, respectively, so we ought to be able to outwit those crea-

tures, if not beat them in fair combat. Unfortunately, though, the mouse comes out better than we do, with a figure of 3.2 percent, but I suppose we can beat that little creature in combat if we can't outwit it. But in small birds, the ratio may be as high as 8 percent. Of course, they have light bodies, which no doubt helps inflate the ratio, but they do seem to have a nasty habit of contradicting every means by which we try to prove human superiority.[48]

Another problem is that, other things being equal, small animals have larger ratios of brain size to body size than do larger animals, so the raw ratio may not be a good index. A more sophisticated approach is to use a statistical technique called linear regression to try to predict brain size from body size. In this way, Harry J. Jerison figured out an index that he termed the *encephalization quotient* (*EQ*), which helps restore a suitable distance between humans and other species.[49] This quotient turns out to be 7.4416 in humans, followed by dolphins at 5.3055, and chimpanzees at 2.4865.[50] The elephant weighs in, as it were, with a quotient of 1.8717, and rats have a miserly quotient of 0.4029. The mouse is blessedly reduced to a quotient of about 0.5, so we can stop worrying about that. This quotient also pretty well gets rid of any serious challenge from birds, although comparison is difficult because the regression formula is rather different for birds. It is just as well, perhaps, that our large brains have given us sophisticated ways to figure out measures that prove that our brains are the largest. There may well be creatures out there busily working on formulae to prove that they, after all, are top dogs.

Some authorities have insisted that the best index of intelligence is the size of the neocortex, and not the brain as a whole. The neortex is the outside layer of the brain, and the most recently evolved, and it also houses the functions that we like to call intellectual. Focusing on the neocortex has the decided advantage of getting rid of birds altogether, since they possess no neocortex at all. Robin Dunbar argued that intelligence is driven by social interaction, so that the larger the social group the greater the need for an enlarged neocortex, simply to cope with all of the social pressures. He then showed that what he called the neocortical ratio, which is the ratio of neocortex to the rest of the brain, increases with the size of the social group.[51] Humans have the largest neocortical ratio, at 4.1, closely followed by the chimpanzee at 3.2. Gorillas lumber in at 2.65, orangutans at 2.99, and gibbons at 2.08. According to the equation relating group size to neocortical ratio, humans should belong to groups of 148, give or take about 50. This is reasonably consistent with the estimated sizes of early Neolithic villages. Of course the modern city confused matters, but if you add up all the people you actually know then a figure of 148 might not be far from the mark.

Fossil evidence shows that brain size remained fairly static in the hominins for some four million years after the split from the apes. For example, *Australopithecus Afarensis*, whose most famous representative is known to the present-day world as Lucy,[52] goes back about 3.5 million years, and had a brain size of about 433 cc, slightly over the chimpanzee size of about 393 cc, but less than that of the much larger gorilla at 465 cc.[53] It was the emergence of the genus *Homo* that signaled the change. *Homo habilis* and *Homo rudolfensis* were still clumsily bipedal but their brains ranged in size from around 500 cc to about 750 cc, a small increase over that of the earlier hominins. *Homo ergaster* emerged a little over 1.8 million years ago, and by 1.2 million years ago boasted a brain size of some 1,250 cc. Thus in a space of about 750,000 years, brain size more than doubled—that's pretty quick on an evolutionary time scale.

Brain size continued to increase at a slower rate. It appears to have reached a peak, not with *Homo sapiens*, dating from about 170,000 years ago, but with the Neandertals, whose fossil remains have been found primarily in western Europe, and as far east as Uzbekistan. Analysis of Neandertal DNA suggests that we share a common ancestor with the Neandertals dating from about 700,000 years ago, and our ancestral populations split from each other about 370,000 years ago,[54] so the increase in brain size may have taken different trajectories. In some individual Neandertals, brain capacity seems to have been as high as 1,800 cc, with an average of around 1,450 cc. Brain size in our own species, *Homo sapiens*, is a little lower, with a present-day average of about 1,350 cc, but still about three times the size expected of an ape with the same body size. This is rather surprising, as we usually congratulate ourselves as being more intelligent than the Neandertals, whom we probably drove to extinction by around 30,000 years ago (as though that were a smart thing to do). Nevertheless, the Neandertals are thought to have had larger bodies than we humans possess, and when brain size is calibrated against body size, we humans may have had the edge. As we shall see below, this final increase in brain size—the dash for the summit, as it were—seems to have coincided with an advance in technological invention over that which had prevailed for the previous 1.5 million years.

We are beginning to learn something of the genetic changes that gave us our swollen heads. One gene known to be a specific regulator of brain size is the *abnormal spindle-like microcephaly associated* (ASPM) gene, and the evidence suggests strong positive selection of this gene in the lineage leading to *Homo sapiens*.[55] Indeed, a selective sweep appears to have occurred as recently as 5,800 years ago, suggesting that the human brain is still undergoing rapid evolution.[56] Another gene known as *microcephalin* (MCPH6) has also been shown to regulate brain size, and one variant in modern humans arose an estimated 37,000 years ago.[57] Other genes involved in the control of brain size that have undergone accelerated rates of protein evolution at points in human lineage have also been identified.[58]

My favorites, though, are two genes that seem to have resulted in increased brain size through negative rather than positive selection; that is, they were once active genes that were inactivated. Moreover, both of them seem to have been inactivated just prior to the time when brain size increased in our own genus, *Homo*, perhaps providing the initial thrust that propelled us toward humanity. One is a gene that encodes an enzyme that produces an acid that inhibits brain growth. This acid is absent in Neandertal fossils and in present-day humans, but present in other primates—although down-regulated in the chimpanzee. It has been estimated that the gene was inactivated around 2.8 million years ago.[59] The other gene encodes the myosin-heavy chain (MYH16) responsible for the strong chewing muscles in most primates, including chimpanzees and gorillas, as well as the early hominins. This gene was

inactivated an estimated 2.4 million years ago, leading to speculation that the shrinkage of jaw muscles and their supporting bone
structure removed a further constraint on brain growth—brawn
gave way to brain.[60] This change may have signaled a change of
diet from tough vegetables to tender meat, or it may have had to
do with the increasing use of the hands rather than the jaws to
prepare food.[61]

Conclusions about the roles of these last two genes in determin-
ing the size and shape of the human brain are of course speculative,
and indeed controversial,[62] but it is nevertheless interesting to reflect
that we may owe our humanity in part to the *loss* of genetic information. Prior to their inactivation, these genes seem to have been
part of a heavenly conspiracy to prevent earthly creatures from
becoming too intelligent. The idea that our humanity may have
depended in part on a reduction in active genes also runs counter
to the intuitive idea that the human mind evolved through the accumulation of new genes, such as the "grammar gene" proposed
by Steven Pinker.[63] One might be tempted to think that some genes
are like taxes, holding back development, and the sooner we get
rid of both the better.[64]

Is It Just a Matter of Size?

Although recursion may have depended partly on the sheer size of
the brain available for computation, some areas are no doubt more
critical than others. One such area is the prefrontal cortex, known
to be involved in planning, and also in mental time travel, as we
saw in chapter 6. It has been claimed, moreover, that the increase
in brain size in humans is disproportionately large in the prefrontal
lobes,[65] although this is somewhat controversial.[66] More detailed
study of the frontal lobes nevertheless suggests that, regardless of
its overall size, the prefrontal cortex may have been reorganized in
the human brain relative to that in the chimpanzee brain. One area
of the human prefrontal cortex is actually smaller than expected
on the basis of overall brain size, but this was at the expense of an
increase in the number of specialized regions in adjacent areas. The
frontal pole of the prefrontal cortex appears to be especially en-

larged in humans relative to apes, especially in the right side of the brain. One article summarizing these findings concludes that these developments have to do with the evolution of "self awareness, social problem solving, the ability to recall personal experiences, and the ability to project oneself into the future."[67] Just what the doctor ordered.[68]

Another critical difference between humans and other primates lies in the way in which the human brain develops from birth to adulthood. We humans appear to be unique among our fellow primates, and perhaps even among the hominins, in passing through four developmental stages—infancy, childhood, juvenility, and adolescence. John L. Locke and Barry Bogin suggest that each of these stages contributes differently to the acquisition of language—see figure 14.[69] During infancy, lasting from birth to age two and a half, infants proceed from babbling to the point that they know that words or gestures have meaning, and can string them together in two-word sequences. This is about the level that the bonobo Kanzi has reached, although we saw in chapter 9 that one-year-old infants point in a way that suggests the sharing of information, whereas chimpanzees do not.

Nevertheless it is the next stage, childhood, that seems to be especially critical to the emergence of grammatical language and theory of mind. According to Locke and Bogin, childhood lasts until about age seven, and is peculiar to humans. They write:

> Childhood is defined by several developmental characteristics, for example, a slowing and stabilization of the rate of growth; immature dentition; feeding characteristics, such as dependence on older people for food; and behavioural characteristics, including immature motor control. The evolutionary value of childhood lies in the mother's freedom to discontinue nursing her three-year-old, which enabled her to initiate a new pregnancy. Doing so enhanced reproductive output without increasing the risk of mortality for the mother, or her infant or older children, for in cooperatively breeding societies others were available to help care for the young.[70]

Childhood seems to be the language link that is missing in great apes and the early hominins, which may account for the fact that, so far at least, great apes have not acquired recursive grammar. But

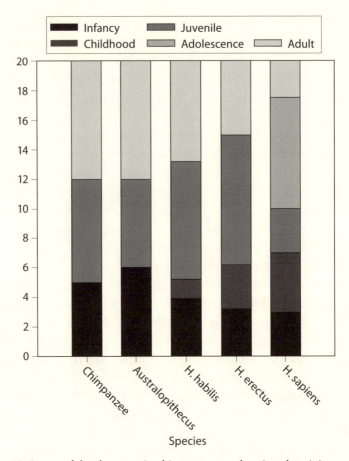

Figure 14. Stages of development in chimpanzees and various hominin species. Note that the childhood stage emerges only in the genus *Homo* (after Locke and Bogin 2006).

it is also during childhood that theory of mind, episodic memory, and understanding of the future emerge.[71] Childhood may be the crucible of the recursive mind.

During the juvenile phase, from age 7 to around 10, children begin to appreciate the more pragmatic use of language, and how to use language to achieve social ends. The final stage is adolescence, which Locke and Bogin suggest is also unique to our own species, and sees the full flowering of pragmatic and social function, in such activities as storytelling, gossip, and sexual maneuvering. Adolescence also has a distinctive effect on male speech, since the surge of testosterone increases the length and mass of the vocal folds, and lowers the vibration frequency. In common parlance, this is the stage when the voice breaks. Chimpanzee parents are perhaps fortunate that their offspring are denied this stage of development.

Locke and Bogin focus on language, but the staged manner in which the brain develops may account more generally for the recursive structure of the human mind. Recursive embedding implies hierarchical structure, involving metacontrol over what is embedded in what, and how many layers of embedding are constructed. Early development may establish basic routines that are later organized in recursive fashion.

Migrations

Although the members of our genus who eventually evolved into *Homo sapiens* remained in Africa, there were evidently migrations both within Africa and out of Africa. Migratory patterns may signal a further step toward humanity. These migrations were not repeated seasonal events, as with many animals and birds, but were perhaps planned, involving the transport of resources and adaptations to new environments. Instead of traveling back and forth between different locations, depending on the season, the migrating hominins tended to move to new locations, and then move on rather than returning.[72]

Remains of the genus *Homo* are found in Asia, where the species is generally known as *Homo erectus*. The nearly equivalent species known as *Homo ergaster* stayed in Africa, but the increase in brain size within Asia seems to have been more rapid than that in Africa.[73] The larger brain and striding gait of *Homo* may therefore have been associated with wanderlust. However, the simple idea that *Homo ergaster* simply wandered out of Africa to become *erectus* in Asia has been complicated by increasing discoveries of morphological and geographic variability within both Africa and Asia. By 1.7 million years ago, there were rather different species of *Homo* in areas as far ranging as southern Africa, equatorial east Africa, Russia, and Java, and stone tool deposits suggest further habitation in Israel, Pakistan, and northern China. This suggests that hominins may have ranged quite widely in the grasslands of

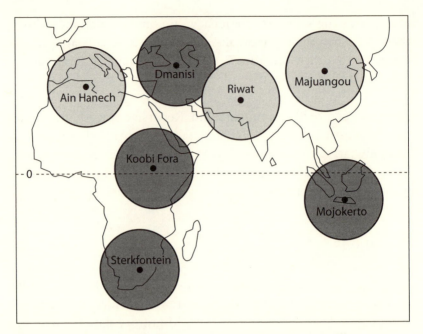

Figure 15. Known or inferred locations of genus *Homo* by 1.7 million years ago. Circles represent radii of 1,000 miles. Dark gray circles represent known populations, light gray circles represent locations where stone tools were made (after Dennell and Roebroeks 2005).

Asia and Africa from as early as 3.0 to 3.5 million years ago, at a time when the Saharan and Arabian deserts did not exist as barriers to migration, and perhaps independently evolved *Homo*-like characteristics.[74] And yet our own species, *Homo sapiens*, was at first more of a stay-at-home, emerging in Africa some 170,000 years ago.

Later, though, it was *Homo sapiens* who exploded out of Africa to eventually populate—one might say overpopulate—the globe. That story will be told in the next chapter.

Tools

Another property once thought to differentiate humans from other primates was the manufacture of tools. We now know that chimpanzee do make simple tools. For example, they fashion sticks for fishing termites out of holes,[75] and make spears for jabbing into the hollow trunks of trees to extract bushbabies, and then eat them.[76] Chimpanzees in the Laongo National Park in Gabon use tool sets comprising up to five different stick and bark tools to extract honey from hives.[77] Some of their tools are composites, such as the use of stones or clubs as hammers to crack nuts on anvils made of stone or wood,[78] or leaf sponges in which several leaves are compressed into an absorbent mass used to extract water from tree holes.[79] These activities do appear to involve some combining of elements, but still have little of the recursive complexity of human manufacture. In a review of evidence on tool manufacture, Benjamin B. Beck wrote, "Unquestionably man [sic] is the only animal that to date has been observed to use a tool to make a tool,"[80] and I know of no evidence to contradict this. In short, we humans—women as well as men, pace Beck—manufacture objects recursively, and it is largely because of our recursive understanding of manufacture that we have polluted the earth with immense cities, not to mention the cobweb embrace of the Internet.

Nevertheless advances in toolmaking were slow. There is little to suggest that the early hominins were any more adept at making or using tools than are present-day chimpanzees, despite being bipedal, and it was not really until the emergence of the genus Homo that toolmaking became more sophisticated. The earliest innovation seems to have been stone tools, with sufficient design features to suggest forward planning, and perhaps the beginnings of mental time travel. The earliest such tools date from about 2.5 million years ago, and are tentatively associated with H. rudolfensis.[81] These tools, relatively crude cutters and scrapers, make up what is known as the Oldowan industry. A somewhat more sophisticated tool industry, known as the Acheulian industry, dates from around 1.6 million years ago in Africa,[82] with bifacial tools and hand-axes. At one time it was thought that these manufactured tools were the distinctive mark of humanity that set us apart from other species, but more recent research has suggested that other species can make tools of similar complexity. As we saw in chapter 6, the champions may be the New Caledonian crows that manufacture tools from pandanus leaves that are precisely tapered for the extraction of grubs from holes.[83] These tools seem on a par with those made by Homo erectus.

The Acheulian industry remained fairly static for about 1.5 million years, and seems to have persisted in at least one human site dating from only 125,000 years ago.[84] Nevertheless at some sites there was an acceleration of technological invention from around 300,000 to 400,000 years ago, when the Acheulian industry gave way to the more versatile Levallois technology. Tools comprising combinations of elements began to appear, including axes, knives and scrapers mounted with hafts or handles, and stone-tipped spears. John F. Hoffecker sees the origins of recursion in these combinatorial tools,[85] which were associated with our own forebears, as well as with the Neandertals, who evolved separately from around 700,000 years ago.[86] An early Neandertal example is the discovery of sophisticated wooden spears, dating from as early as 400,000 years ago, that were preserved in a coal mine in Schoningen in Germany. They were associated with the fossil remains of horses,[87] and suggest an advanced hunting technology.[88] The hominins who inhabited this site were Neandertals (or their predecessors), not the predecessors of *Homo sapiens*.

Tools are of course important to the human story, but there is little evidence that they were decisive in creating the human mind. To be sure, recursive elements were evident in tools from half a million or so years ago, but a truly manufactured world did not really emerge until after the appearance of *Homo sapiens*, and varies widely between different cultures. My guess is that recursive thought probably evolved in social interaction and communication before it was evident in the material creations of our forebears. The recursiveness and generativity of technology, and of such modern artifacts as mathematics, computers, machines, cities, art, and music, probably owe their origins to the complexities of social interaction and storytelling, rather than to the crafting of tools. I have more to say about technology in chapter 12.

Human at Last

It is generally reckoned that the species *Homo sapiens* who emerged some 170,000 years ago was "anatomically modern"—human at last. This was a large-brained species equipped with the intelligence and social understanding of modern humans. If you were to snatch an infant from that time, and raise that infant in the present-day Western world, he or she would probably adapt as well as any modern-born person to the exigencies of modern life, whether as stockbroker, ballerina, modern-day hunter-gatherer, university professor, or used-car salesperson. The Pleistocene had gradually shaped recursive modes of thought that allowed complex theory of mind and mental time travel, and allowed for the relaying of memories, plans, and stories for the betterment of both the society and the individual.

But perhaps this scenario is not quite right. Human evolution was to go through another phase that was to transform society and create the vast complexities of civilization. It remains a matter of contention whether this involved biological change, but it was nevertheless sufficient to see the extinction of that other large-brained species, the Neandertals. In other words, *Homo sapiens* had become human, but had not yet become modern, with all of the benefits and perils that term implies.

To explain it, we need another chapter.

12

Becoming Modern

One of the few good things about modern times: If you die horribly on television, you will not have died in vain. You will have entertained us.
   —Kurt Vonnegut, *Cold Turkey* (2004)

Homo sapiens emerged in Africa about halfway through the period known as the Middle Stone Age, which began around 300,000 years ago and ended around 50,000 years ago. Early *sapiens* may have been anatomically modern, but in terms of culture and technology was probably not greatly distinguishable from other large-brained members of the genus *Homo*. These included the Neandertals, who died out in Europe some 30,000 years ago, apparently eclipsed by the arrival, some 20,000 years earlier, of our own predatory species.[1] We do not know of course what the Neandertals might have achieved had they survived, but *Homo sapiens* brought a technological and cultural sophistication apparently unmatched by our hapless Neandertal cousins. For the origins of this flowering of what we are pleased to call modernity, we need to look to Africa, whence our species came.

   Understanding of the different lineages of *Homo sapiens* has come from studies of mitochondrial DNA (mtDNA), a form of DNA that is passed down the generations from mother to daughter. Unlike nuclear DNA, it is not involved in the processes of recombination that occur in eukaryotic reproduction. Changes in mtDNA arise only through mutation, allowing us to trace ancestry through females, uncontaminated by interference from us mere males. Analysis of mtDNA has identified four lineages in Africa prior to the exodus of some members of the species some 60,000 years ago.[2] These lineages, known as *haplogroups*, are imaginatively labeled L0, L1, L2, and L3. Routines are now available for estimating the

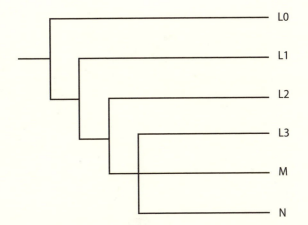

Figure 16. Branching of the early lineages (*haplogroups*) of *Homo sapiens* (after Atkinson, Gray, and Drummond 2009).

population sizes of different lineages, and tracking them through time. These lineages, along with two others, M and N, which are found out of Africa, appear to have split off as shown in figure 16.

The population of the earliest lineage, L0, is estimated to have expanded through the period 200,000 to 100,000 years ago. This lineage presumably included the hypothetical individual known as "mitochondrial Eve,"[3] the mother of us all, so to speak. Present-day descendants include the Khoisan of southwest Africa, speakers of the click languages that were exploited to great effect by the singer Miriam Makeba. The L0 and L1 lineages exist at higher frequencies than the other lineages among present-day hunter-gatherers, who may therefore offer a window into the early history of *Homo sapiens*.

Out of Africa

The L3 lineage is of special interest, because it expanded rapidly in size from about 60,000 to 80,000 years ago, and seems to have been the launching pad for the migrations out of Africa that eventually populated the globe. Of the two non-African lineages that are the immediate descendants of L3, lineage M is estimated to

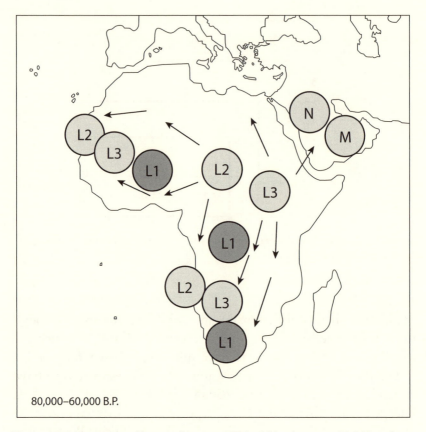

Figure 17. Movements of human lineages within Africa between 80,000 and 60,000 years ago (after Mellars 2006b).

have migrated out of Africa between 53,000 and 69,000 years ago, and lineage N between 50,000 and 64,000 years ago.[4]

Why did L3 expand so rapidly, and migrate from Africa? One suggestion is that L3 gained some cultural advantage over the other lineages, perhaps through the invention of superior technologies, and that this gave them the means to migrate successfully. Paul Mellars suggests that the African exodus was predated by advances in toolmaking, including new stone-blade technologies, the working of animal skins, hafted implements, and ornaments.[5] Some of the improvements in tool technology can be attributed to the use of fire to improve the flaking properties of stone, which dates from around 72,000 years ago on the south coast of Africa.[6] The use of fire for cooking, and perhaps for warmth and protection, probably goes back about 800,000 years,[7] but its use in technological development suggests enhanced inventiveness and cognitive understanding. In excavations at Blombos Cave on the Southern Cape coast of South Africa, Christopher S. Henshilwood has unearthed ochre bars with carved abstract designs, and shell beads serving as personal ornaments.[8] These and other artifacts from Blombos Cave and neighboring sites in South Africa are said to somewhat resemble much later artifacts in Europe, and may represent technologies exported by L3.[9]

It need not follow that the L3 people were biologically more advanced than their African cousins, and it may well be that the exodus was driven by climate change rather than any technical superiority of L3 over the other haplogroups that remained in Africa. During the last ice age, there were a series of rapid climate swings known as Heinrich events. One of these events, known as H9, seems to have occurred at the time of the exodus from Africa, and was characterized by cooling and loss of vegetation, making large parts of North, West, and East Africa inhospitable for human occupation. It may also have been accompanied by a drop in sea levels, creating a land bridge into the Levant.[10] So out of Africa they went, looking no doubt for greener pastures.

The exodus seems to have proceeded along the coast of the Red Sea, across the land bridge, and then round the southern coasts of Asia and southeast Asia, to reach New Guinea and Australia by at least 45,000 years ago. Mellars notes similarities in artifacts along that route as far as India, but remarks that technology seems to have declined east of India, especially in Australia and New Guinea. This may be attributable, he suggests, to the lack of suitable materials, adaptation to a more coastal environment requiring different technologies, and random fluctuations (cultural drift). One remarkable point of similarity, though, is the presence of red ochre in both Africa and in the earliest known human remains in Australia.[11] Ochre was probably used in ritualistic body-painting, and perhaps in painting other surfaces.[12]

Precisely how the dispersal reached Europe is not yet entirely clear. One notion is that a separate dispersal from Africa proceeded up the Nile Valley, but Mellars suggests instead it was the single southern migration that eventually split, with one arm proceeding to New Guinea and Australia, and the other proceeding northward through Arabia or Iran to reach Europe and the Near East by 40,000 to 45,000 years ago.[13]

It is commonly suggested that the relatively sudden spurt of manufacture and symbolic behavior in the Middle Stone Age, and perhaps the exodus from Africa, signaled the emergence of language. In chapter 2 we encountered Chomsky's notion that language emerged through some fortuitous mutation in a single individual, whom Chomsky named Prometheus. Timothy Crow has endowed Prometheus with the honor of initiating our species, with all of the characteristics that single us out from the other hominins.[14] As we saw in chapter 4, others have suggested that the critical mutation that gave us language occurred even later, perhaps as recently as 50,000 years ago, and this led in turn to the remarkable rise of sophisticated technology. John F. Hoffecker writes:

> Modern human technological ability seems to be an integral part of a wider package of behavior ("behavioral modernity") that developed in the context of the African MSA [Middle Stone Age] before 50 ka. Modern human technology exhibits many of the characteristics, most notably the creativity and structural complexity, of art, music, ornament, and other forms of symbolism (*including by implication syntactical language*) that are elements of behavioral modernity. Modern humans, as they dispersed out of Africa, adapted quickly to a wide range of habitats designed during the late African MSA or created in response to local conditions.[15]

But was it really *language*? In chapter 4 I argued that language evolved first as a system of manual gestures, shifting gradually through facial gestures to articulate speech. In this view, language itself evolved well before the emergence of *sapiens*, a product of the Pleistocene rather than specifically of our own species. It may be significant that click languages are associated with the descendants of L0, and may be residues of prevocal language. Click sounds are made entirely in the mouth, and some languages have as many as 48 click sounds.[16] Clicks could therefore have provided sufficient variety to carry a form of language prior to the incorporation of vocalization. Two of the many present-day African groups that make extensive use of click sounds are the Hadzabe and San, who are separated geographically by some 2,000 kilometers, and genetic evidence suggests that the most recent common ancestor of these groups goes as far as 100,000 years ago.[17]

My guess is that gestures and perhaps clicks gave way to a more effective sound-based language at some point prior to the exodus from Africa. Indeed, some of the phonemes we use, as represented by letters such as /k/, /p/, and /t/, are not voiced, and may be the remnants of earlier click sounds. The incorporation of voicing may have involved a mutation of the *FOXP2* gene, but as I explained in chapter 4, there is still uncertainty about precisely when the most recent mutation occurred, or whether it was truly critical to the evolution of speech. Alternatively, the switch to vocal speech may have been a cultural invention, adopted because of its practical advantages over manual forms of communication.[18] And of course we still gesture as we speak,[19] and children readily learn signed languages where speech is prevented.

But we don't have to wave our arms about in order to communicate effectively, as illustrated by radio and the ubiquitous cell-phone. The emergence of speech may therefore have freed the hands from obligatory use of manual gestures to communicate. Perhaps this newfound freedom enhanced transportation, so the hand could be more readily used for carrying belongings in the migrations from Africa, or in seeking more hospitable regions within Africa. The hands might also have been freed for the more effective use and manufacture of tools. At the same time, speech could be employed to explain manufacturing techniques, as illustrated by cooking shows on television. In short, my suggestion is that the emergence of modernity, and the flowering of manufacture and the other embellishments of modern humans, originated not with language but with the emergence of speech.[20]

The African record prior to the exodus certainly suggests the beginnings of modernity, although the development of technology and cultural complexity seems relatively meager compared with what was to come in the Upper Paleolithic, or Late Stone Age, which is generally reckoned to have ranged from some 40,000 years ago to 12,000 years ago.

The Upper Paleolithic

This second wave of innovation was most pronounced in Europe and western Asia, beginning roughly when *Homo sapiens* arrived there. The Upper Paleolithic marked nearly 30,000 years of almost constant change, culminating in a level of modernity equivalent to that of many present-day indigenous peoples. Technological advances included clothing, watercraft, heated shelters, refrigerated storage pits, and bows and arrows. Elegant flutes made from bone and ivory have been unearthed in southwest Germany, dated at some 40,000 years ago,[21] suggesting early musical ensembles, if not rock concerts. Flax fibers dating from 30,000 years ago have been found in a cave in Georgia, and were probably used in hafting axes and spears, and perhaps to make textiles; and the presence of hair suggests also that they were used to sew clothes out of animal skins.[22] The people of this period mixed chemical compounds, made kilns to fire ceramics, and domesticated other species. Stone tools date from over two million years ago, but remained fairly static until the Upper Paleolithic, when they developed to include more sophisticated blade tools, as well as burins and tools for grinding. Tools were also fashioned from other materials, such as bone and ivory, and included needles, awls, drills, and fishhooks.[23]

Vanity seems to have evolved even further during the Upper Paleolithic, as people began to decorate themselves not only with ochre, but also with threaded shell beads, tattoo kits, and other forms of personal ornamentation. Art became more sophisticated, as illustrated by the cave drawings of large animals in various regions of Europe, especially France and Spain. The most spectacular is the Chauvet Cave in southeastern France, with lifelike drawings of horses, bison, deer, rhinos, and cave lions. Figurines dating from over 30,000 years ago in Germany represent the first appearance of fully rounded sculptures. The earliest-known is a sexually explicit figurine of a woman, dating from 35,000 years ago, recently

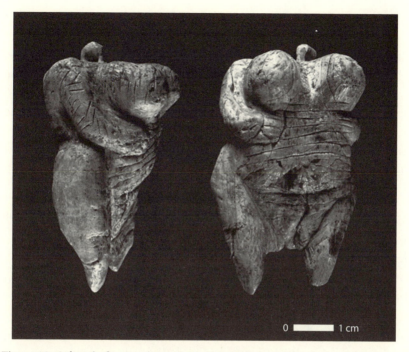

Figure 18. A female figurine from Hohle Fels Cave in southwestern Germany. (Reprinted by permission from Macmillan Publishers Ltd: *Nature*, Conard 2009.)

discovered in a cave in south Germany. This remarkable sculpture shows exaggerated sexual features, including enlarged breasts, an enlarged and explicit vulva, and bloated belly and thighs. In this instance, at least, sex may have reared its ugly body but not its head, since the figurine has no head; in its place is a carved ring, suggesting that the figurine was worn as a pendant.[24] This is early evidence, you may cynically think, of the female body treated as a sexual object, but a bone carving of an erect phallus found in southwestern France dating from around the same time restores some balance. Paul Mellars writes that one of his students suggested that this and other phallic objects might have been girls' toys.[25]

The most dramatic event of the Upper Paleolithic, though, was the extinction of the Neandertals, who had long inhabited Europe and western Asia before *Homo sapiens*, in the form of Cro-Magnon humans, arrived some 45,000 years ago.[26] Whether the Cro-Magnons simply adapted better to the ecology of the region or exterminated the Neandertals by more direct means is unknown, although the human record of warfare and genocide rather suggests the worse. Humans may well have wiped out the Neandertals in much the same way as they have wiped out whole communities of their own species, such as the aboriginal people of Tasmania. Genocide is an all too frequent escalation of intertribal conflict. Darwin even argued that such conflict was itself an important contributor to biological evolution:

> Natural selection arising from the competition of tribe with tribe . . . would, under favourable conditions, have sufficed to raise man to his present high position in the organic scale.[27]

Hmmm.

But the Upper Paleolithic did not only enhance the capacity to kill. To some, it was the Upper Paleolithic, not the Middle Stone Age, that marked the transition to modernity, with its positive as well as negative aspects. Developments during this era have been said to comprise a "human revolution,"[28] marking the true discontinuity in the evolution of humankind. The anthropologist Richard G. Klein, for instance, has argued that the disparity in artifacts between the Upper Paleolithic and earlier developments in Africa, such as those at Blombos Cave described above, were such that some genetic mutation might have taken place.[29] Only in the Upper Paleolithic, he thinks, did we emerge as truly modern humans, biologically as well as culturally.

Again, then, we see an appeal to a sudden, fortuitous mutation as the key to human uniqueness—a second coming of Prometheus, perhaps. But the apparent disparity between developments of the Middle Stone Age and those of the Upper Paleolithic probably depended more on demography than on any genetic change. One important factor may have been geographically local increases in subpopulation sizes, heightening the competition for resources and driving more complex social organization. We saw earlier that the haplogroup L3 underwent an expansion from about 60,000 to 80,000 years ago that marked an increase in cultural and technological complexity, and prompted the migration from Africa. Migration into new territories would itself have increased the pressure for the invention of new technologies, including those necessary for adjusting to changes in climate and available resources. Migration can itself bring about rapid change in the size of a population, as migrants descend on the same territory, adding to the pressure for cultural innovation. It has been estimated that the density of the human population in Europe increased rapidly from the first arrival of the Cro-Magnons, and by around 45,000 years ago had reached the same density of population as that of sub-Saharan Africa about 100,000 years ago.[30]

It is still not altogether clear, though, why there was a gap in the appearance of "modern" traits between 70,000 years ago in Africa and their appearance 40,000 years ago in Europe. Perhaps people actually on the move had less opportunity for innovation, or perhaps the artifacts they left are more scattered and less easy to find. And depopulation can have a reverse effect, leading people to lead simpler lives, and maybe even happier ones. Migration out of Africa might therefore have reduced population size, both among those who remained in Africa and among the bands that dispersed along the southern Asian coasts and into New Guinea and Australia, and north into Europe. But populations at various endpoints might then have built up, explaining bursts of new technology. Another factor must have been the catastrophic eruption of Mt. Toba in Sumatra around 70,000 years ago, which plunged the world into what has been described as a "volcanic winter," or mini-ice age, which lasted several thousand years and probably reduced the world's human population to around 10,000.[31] We must hope that global warming doesn't have a similar effect—and of course another calamitous eruption could occur at any time.[32]

The human condition is in any case one of extraordinary variability in both time and space, governed more by ecology than by physiology.

Beyond the Stone Age

During the Stone Age, people were hunter-gatherers. Beginning perhaps 13,000 years ago, at the end of the last Ice Age, diversification increased markedly. People began to domesticate wild plants and animals for their own purpose. One of the cradles of domestication was the Fertile Crescent, so-called because of its fertile soil and half-moon shape. It corresponds broadly to present-day Iraq, Syria, Lebanon, Israel, Kuwait, Jordan, southeastern Turkey and southwestern Iran. Other areas of domestication were China, Mesoamerica, and what is now the southeastern United States, and each of these regions in turn developed different agricultural and industrial technologies. Other areas of the world, such as parts of Africa, New Guinea, and Australia, retained a hunter-gatherer existence. Jared Diamond's book *Guns, Germs, and Steel* attributes the vast differences between peoples to the exigencies of history and geography.

Diversity is probably exacerbated, though, by the tendency of groups, or tribes, to actively create distinctiveness. Within-group bonding is achieved in part by creating between-group difference, so that outsiders or freeloaders are kept out. Languages, for example, evolve to the point of being mutually inaccessible, and I suggested in chapter 4 that one of the functions of language, especially in the form of speech, is to create impenetrable social fortresses. The bonding of peoples within those fortresses, and the separation of fortresses from each other, creates further diversity. People in different areas of the world also develop different habits and skills. Culture is as much a mechanism for bonding as a mechanism for intercultural divergence, maintained through such institutions as religion, language, and custom, whether based on hunting and gathering, agriculture, industry, or trading. And sport, for that matter. One of the challenges of moving between countries that share even a quite recent cultural heritage is to understand and enjoy the local sporting obsessions, be it ice hockey in Canada, baseball in the United States, or Australian Rules—a form of football, I think. And not all nations, I discover, share the New Zealand obsession with rugby.

Cultural diversity is especially evident in technology, which also provides superb examples of the recursive nature of human endeavor, with wondrous inventions ranging from internal combustion to the Internet, railways to rockets, lasers to laptops, eyeglasses to iPods. The economist W. Brian Arthur points out that technological invention consists of the hierarchical assemblies and subassemblies, combined in recursive fashion. Here is his description of a jet engine:

> A jet engine has a main assembly that consists of an air intake system, a compressor system (to compress the inducted air for combustion), a combustion system (to provide high-energy gas flow for the turbine), a turbine system (to drive the compressor and provide reactive thrust), and an exhaust section. Each of these in turn is controlled, supplied, and monitored by other subsystems: the compressor system requires a variable vane actuating system (to set the vane angles appropriate to airflow velocity), and an anti-stall bleed system to control pressure surges (the tendency of the compressed air to blow backwards); the turbine system requires a blade cooling system, and a complicated set of shrouds and seals to prevent high-pressure gas leakage.[33]

And of course the jet engine is just a part of the larger assembly— the jet aircraft itself.

One might be tempted to see such intricate technology as the crowning achievement of the recursive mind, creating an unlimited set of mechanical possibilities. It is as though the mind has spilled over to inhabit the earth, and indeed to control it. In its recursive, combinatorial nature, technology is like language—and indeed language itself might be regarded as a technological solution to the problem of communication. Unlike language, though, technology is not universal. Some cultures exploit technology to the point of strangling the life out of the planet, and others retain relatively primitive technologies of the Stone Age. More importantly, technology is not so much the achievement of the individual human mind as the cumulative product of culture over many generations. The great British scientist Sir Isaac Newton wrote, "If I have seen further than others it is only by standing on the shoulders of giants."[34] Most of us would have no idea how to build a jumbo jet or a television set. These things have emerged through the ratchet-like advances of culture.[35] We have adapted to them, but doing so is surely no more sophisticated than the way aboriginal Australians have adapted to the subtle cues that allow them to track down animals, or to find and identify edible plants.

Diversity does not preclude the mutual recognition of underlying humanity. Daniel Everett, in his time among the Pirahã in Brazil, met a culture as remote from his own as it is possible to find. Their language, religion, and mode of life were utterly different from anything he had previously experienced in the United States, and yet he eventually learned their language, even if he failed to convert them to his religion. He also developed lasting friendships. Jared Diamond, an American, formed similar bonds with the native people of New Guinea, where he worked for a time. He records his friendship with Yali, a local politician. Although the white colonialists of New Guinea regarded the native people there as "primitive," it was clear to Diamond that they were at least as intelligent as Europeans, and in Yali in particular Diamond recognized an intellect he considered superior to his own. Yali had a question for Diamond: "Why is it that you white people developed so much cargo and brought it to New Guinea, but we black people had little cargo of our own?"[36]

The vast differences in cargo between the people of New Guinea and those of modern European, North American, or Australasian society can only reinforce my belief that the essence of humanity is not the things we make, but rather the way we think. Our recursive understanding of each other, and our recursive ability to tell stories, whether fictional or autobiographical, is what truly sets us apart from other species, but aligns us with our fellow humans, of whatever race or culture.

13

Final Thoughts

It was a bright cold day in April, and the clocks were striking thirteen.

—George Orwell, *Nineteen Eighty-Four*

So began Orwell's famous novel. Thirteen is a beautiful harmonious number. You can fit 12 equal-sized spheres around a central sphere, making thirteen in all, and now is the time to return to that central sphere. But my recursive mind thinks it knows what you're probably thinking: Chapter 13 of Darwin's book *The Descent of Man* is for the birds,[1] and the thirteenth Tarot card is Death.

Charles Darwin wrote:

> The difference in mind between man and the higher animals, great as it is, certainly is one of degree and not of kind. We have seen that the senses and the intuitions, the various emotions and faculties, such as love, memory, attention, curiosity, imitation, reason, &c, of which man boasts, may be found in an incipient, or even sometimes in a well-developed condition, in the lower animals.[2]

In this book, I have tried to argue that recursion holds the key to that difference in mind, underlying such uniquely human characteristics as language, theory of mind, and mental time travel. It was not so much a new faculty, though, as an extension of existing faculties. Animals communicate, but in the course of human evolution recursive principles were added, allowing us to construct and understand an unlimited number of possible messages. Animals may have some awareness of the mental states of others, but the recursive principle extends this so that one may know that others know what one knows, providing for extended empathy and co-operation. It also introduces enhanced opportunities for deception and exploitation, which drove theory of mind to deeper levels of recursion and Machiavellian intrigue. Animals have memory, but our forebears added a recursive principle that allowed them to insert past episodes into present consciousness, and generate potential future episodes. The generation of episodes led also to storytelling and fiction. I have suggested in this book that the recursive principle evolved in these facets of social life, allowing our forebears to bond, and to share information within social groups, but also to compete, and to indulge in ever-escalating forms of warfare between groups.

I have argued that it was the special conditions of Pleistocene that drove these recursion-based extensions to the mind. Our primate predecessors had adapted to forested environments, and the relative safety of the forest canopy. Indeed, we still bear some of the bodily attributes of that adaptation, such as long arms and grasping hands, and a shoulder joint that still allows us to swing from a horizontal bar—or branch. And it is somehow comforting for the well-heeled to live in leafy suburbs, if not the urban jungle. The challenges that the open and dangerous savanna posed to a species that had adapted to a forested environment forced our hominin forebears into the "cognitive niche," dependent on mutual support and communication, and attention to the microstructure of human behavior and interaction.

Although recursion was critical to the evolution of the human mind, I argued in chapter 1 that it is not a "module," the name given to specific, innate functional units, many of which are said to have evolved during the Pleistocene. Nor did it depend on some specific mutation, or special kind of neuron, or the sudden appearance of a new brain structure. Rather, recursion probably evolved through progressive increases in short-term memory and capacity for hierarchical organization. These in turn were probably dependent on brain size, which increased incrementally, albeit rapidly, during the Pleistocene. But incremental changes can lead to sudden more substantial jumps, as when water boils or a balloon pops. In mathematics, such sudden shifts are known as *catastrophes*,[3] so we may perhaps conclude that emergence of the human mind was catastrophic.

The jump to recursion is perhaps best illustrated by the human ability to count. Many animals are capable of counting, at least in the sense of being able to distinguish between different numbers of objects, or of being able to give different labels to different numbers of objects. As I mentioned in chapter 3, the parrot Alex, late companion of Irene Pepperberg, could count up to six, using the actual spoken labels "one" to "six" to enumerate the number of objects—regardless of their identities. But humans, at least in literate societies, can count up to any number, using the recursive principle articulated in chapter 1. Adding this principle produces what seems a disproportionate leap forward, and may come as a revelation. A realization, an "Aha!" experience, a flash of insight, the glimpse of infinite possibility—these are the building blocks of the human mind. And yet they derive from incremental changes.

It is sometimes suggested that technology is the key to human evolution.[4] To be sure, the marvels of modern technology seem to be the most distinctive marks of humanity on the planet. No other species has so transformed the physical environment, almost to the point of self-destruction. Recursive technology was certainly critical in the evolution of modernity, and indeed postmodernity, but it is not really a human universal. It varies hugely across different peoples on the planet—some cultures have largely retained a hunter-gatherer style of living, with relatively undeveloped technologies, but they are nevertheless fully modern with respect to language and social intelligence. This suggests that technology was a relatively late application of recursive principles, and not an obligate one. Language, in contrast, is a human universal, and I suspect that theory of mind and mental time travel are too.[5]

I have argued in this book that the extension of recursive principles to manufacture and technology was made possible largely through changes in the way we communicate. Language evolved initially for the sharing of social and episodic information, and depended at first on mime, using bodily movements to convey meaning. Through conventionalization, communication became less mimetic and more abstract. In the course of time it retreated into the face, and eventually into the mouth, as late *Homo* gained voluntary control over voicing and the vocal tract, and the recursive ability to create infinite meaning through combinations of articulate sounds. This was an exercise in miniaturization, releasing the rest of the body, as well as recursive principles, for manipulation of the physical environment.

It was this release of bodily resources, I think, that underpinned the development of technology, the cumulative application of recursive principles to material construction. Although simple combinations of elements were established well before *Homo sapiens* emerged, in the form of multielement hand axes and spears, truly recursive technology was much later, and was much more evident in some societies than in others. Thus technology was not the driving force of recursion, but was rather a later discovery of the power of recursion in manufacture. But once recursive principles were introduced into technological invention, their power was immense. W. Brian Arthur calculates that hierarchies in modern technology can be five or six layers deep[6]—and in chapter 8 I noted Robin Dunbar's view that a depth of about five is what is necessary to find God.[7]

The switch from hand to mouth was perhaps the first of a series of changes in the mode of communication to have a dramatic influence on human affairs. The later invention of writing greatly enhanced the storage, accumulation, and sharing of knowledge. Libraries contain vastly more information than any single brain can hold. Written language also led to mathematics, ways of expressing complex relations that in turn led to more advanced calculation, and further technology. The invention of the telephone, radio, television, and Internet are all communicative devices with obvious impact in the creation of a global society. Google makes limitless knowledge available with a few clicks of a mouse, so that libraries—and perhaps the human mind itself—are threatened with redundancy.

The complexities of the modern world are not of course the products of individual minds. Rather, they are the cumulative products of culture. Most of us have no idea how a jet engine, or a computer, or even a lightbulb, actually works.[8] We all stand on the shoulders of giants. The combination of recursive principles and cumulative culture is indeed a powerful one. Thus simple gossip has evolved into the novel and apparently endless soap operas, simple tools into complex technology, simple dwellings into vast cities, rhythmic drumming into symphonies, counting into mathematics. Fermat's last theorem, based on Fermat's conjecture in 1637 that $x^n + y^n = z^n$, has no integer solution for $n > 2$, was not proven until 1995. Andrew Wiles took several years to construct the proof, which depended on a number of other prior mathematical developments, each in itself a maze of recursive complexity, and Wiles's proof still required a couple of articles for full presentation.[9]

And of course we don't all build our minds in the same way. People may be mechanics, poets, musicians, accountants, shamans, lawyers, or authors silly enough to try to explain the human mind. In the terminology of William D. Hamilton,[10] people people veer to the more social aspects of recursive thought, things people toward the mechanical. But recursive principles, once unleashed, create possibilities beyond those that any single human mind can accommodate. It is sometimes said that we use only 10 percent of our brains, a notion roundly debunked by the late Barry L. Beyerstein.[11] Yet it is also true that there are many feats of skill or intellect that we could achieve if only we put our minds to it, as it were. I regret not learning to play the piano or ski,[12] yet I'm sure I could have done these things had I set out to do so at an earlier age. But I suppose there is simply not the time or the motivation, or even the brain capacity, for any individual to develop more than a tiny fraction of the possibilities available from birth. In this respect, at least, humans are better understood as generalists than as bundles of specialized modules. And if I had taken up skiing, you might have been spared this book.

Coda

The driving force in our understanding of human language in the latter part of the twentieth century was Noam Chomsky, who saw that the recursive nature of language distinguished it from other forms of animal communication. Chomsky also understood that language depended on recursive thought, which he named I-language. To Chomsky, though, I-language remains mysterious and little understood, and was so deeply buried in the mind, without reference to the world, that it could not have evolved through natural selection. Instead, he argued, it must have emerged in a single, all-or-none step. Further, this singular event must have occurred in a member of our own species, *Homo sapiens*, perhaps within the last 100,000 years.

In this book, I hope I have convinced you that we know much more about the nature of human thought than is incorporated in the concept of I-language. In this book I have focused primarily on human imagination—especially in the form of mental time travel—and theory of mind. These processes make use of recursive principles, and open up the human mind to endless possibilities. They are not fundamentally linguistic, as Chomsky assumed I-language to be; rather, it was language that adapted to thought, and not thought that was shaped by language. Finally, and critically, there is no reason to suppose that the recursive mind evolved in some single, miraculous step, or even that it was confined to our species. Instead, it was shaped by natural selection, probably largely during the last two million years.

Notes

CHAPTER 1

What Is Recursion?

1. For a useful discussion, see Fitch (2010b), who identifies three different meanings of the word "recursion."

2. From Littlewood 1960, 40.

3. *Lost in the Funhouse*, by John Barth (1969).

4. This was actually a parody on the earlier poem by the seventeenth-century writer Jonathan Swift:

> So nat'ralists observe, a flea
> Hath smaller fleas that on him prey,
> And these have smaller fleas that bite 'em,
> And so proceed *ad infinitum*

My favorite, though, is Ogden Nash's flea-like poem, said to be the shortest ever written:

> Adam
> Had 'em.

5. In case you're old-fashioned, I use * rather than × to mean "multiplied by."

6. Some readers may recognize the Fibonacci series from Dan Brown's book *The Da Vinci Code*, where the first eight digits of the series provided the number of a critical bank account.

7. Pinker and Jackendoff 2005, 203.

8. Fitch 2010a.

9. Pinker and Jackendoff 2005, 230.

10. A case in point is a claim that starlings can parse recursive sequences of sounds (Gentner et al. 2006), leading to a headline in the *New York Times* of 2 May 2006 that read, "Starlings may Shed Light on Language." It turns out that there is a simpler explanation for what the starlings were able to achieve. This is discussed more fully in chapter 3.

11. Chomsky 1995.

12. The champion may be an African grey parrot, who could count accurately up to about six (Pepperberg 2000). This kind of counting is known as *subitization* (Kaufman et al. 1949), as opposed to the iterative/recursive counting that allows us to count indefinitely.

13. Chomsky (1988, 256) wrote that "human language has the extremely unusual, possibly unique property of discrete infinity."

14. This is recursive embedding of mental states, in that Sir Toby's anticipation is embedded in what Maria foresees, Olivia's judgment is embedded in what Sir Toby anticipates, and so on.

15. You can find it, needless to say, on the Web.

16. From Saki's story "The hounds of fate"—see Saki 1936.

17. Hennessy 1995.

18. In some cases, iterative processes can achieve the same effect as recursive ones. Earlier, I showed how recursion can be used to generate infinite mathematical series, such as the natural numbers or the Fibonacci series, but the same series can be generated iteratively. To generate the natural numbers, for example, one can simply program an instruction that might read as follows:

Define function **successor** [while $i > 0$: print $i + 1$].

When the number 1 is entered, the string of integers will be printed forever, or the printer wears out—or someone kindly switches it off. Similarly, one might generate the Fibonacci series as follows:

Define function **fibonacci** [while $i > 1$: print $\{(i - 1)+(i - 2)\}$].

These definitions are not themselves recursive in that the function does not call on itself, but they can operate indefinitely if each output is fed back into the function as the next input.

19. An example familiar to psychologists is factor analysis. The problem is to estimate so-called communalities, which are the (unknown) diagonal elements in a correlation matrix. You start by guessing what the communalities might be, compute a factor solution, from which new communalities are computed. You then repeat the process until the communalities stabilize. As a punishment for missing labs, I was once forced to compute a solution by hand. It took hours, but those were the days before computers.

20. The approach also owed much to Jerry Fodor's (1983) book *The Modularity of Mind*, although Fodor himself, in response to Pinker's *How the Mind Works*, wrote a book entitled *The Mind Doesn't Work That Way*. Part of his objection seems to be to what he calls the "massive modularity hypothesis," which is the idea that the whole damn thing is modular, and part to the incorporation of modules into a Darwinian view of mental evolution. I use the term "evolutionary psychology" to refer to the group of psychologists who adopt the basic tenets laid out by Barkow, Cosmides, and Tooby (1992). Other psychologists, such as myself, are interested in evolution, but don't necessarily go along with all of these tenets.

21. Pinker 1997, 27.

22. Human evolution during the Pleistocene is further discussed in chapter 11, although not specifically from the perspective of the evolutionary psychologists.

23. One of the main proponents was William McDougall (1908).

24. Bernard 1924.

25. Pinker 1997, 315.

26. Cosmides 1985.

27. New, Cosmides, and Tooby 2007.

28. Quoted in *Science*, **318**, 25 (2007).

29. Pinker 1997, 315.

30. Mithen 1996.

31. Premack 2007, 13866.

32. Read 2008. It is not clear whether the limitation is one of simple capacity, or whether chimpanzees lack the specific ability to store embeddings.

33. Throughout the book, I use the term "hominin" to refer to humans and their extinct bipedal ancestors, but not including chimpanzees and the other great apes. This is consistent with most current usage (e.g., Wood and Collard 1999), although some authors retain the term "hominid" for humans and their ancestry, and others include the great apes among the hominins.

PART 1

Language

1. Christiansen and Kirby 2003, 1.

CHAPTER 2

Language and Recursion

1. Müller 1873.
2. Butler 1919, 95.
3. Chomsky 1957.
4. Hauser, Chomsky, and Fitch 2002.
5. Karlsson 2007. I found no instances of three, or even two, levels of center-embedding in his article.
6. Ribena is a black-currant drink with a high concentration of vitamin C. It was heavily promoted in Britain during World War II, when other sources of Vitamin C were scarce.
7. A rewrite rule can be interpreted to mean "replace the expression on the left with the expression on the right." I'm told that rewrite rules are now considered by linguists to be an old-fashioned way of showing the structure of sentences, but here they make the point.
8. Chomsky 1975. For a useful summary, see also Chomsky 2010, which also includes his views on the evolution of language.
9. Chomsky 2010, 59.
10. Crow 2010. See also note 4 in chapter 4.
11. Noam Chomsky, quoted in Piattelli-Palmarini 1980, 48.
12. Chomsky's terminology has changed—one might even say evolved—over the years. In his early work he contrasted surface structure with *deep structure*. Deep structure seems to have given way to *universal grammar*, and more recently to *I-language*. Needless to say, these changes are accompanied by subtle changes in meaning.
13. Christiansen and Chater 2008.
14. Life among the Pirahã proved physically hazardous as well as linguistically impenetrable to a Westerner—see Everett's book *Don't Sleep, There are Snakes* (2008).
15. Everett 2005.
16. Everett 2005, 629.

17. Everett 2005, 634.

18. See the critique by Nevins, Pesetsky, and Rodrigues (2007) and the response by Everett (2007). The issue may depend on how recursion is defined. The Pirahã language may have no recursion in the sense of embedded phrases, but may still be recursive in the sense of Chomsky's Merge operation. In the example of *My saying John intend-leaves*, the phrase *My saying* and *intend-leaves* might each be considered products of Merge, which are then themselves merged into the utterance (not strictly a sentence as it contains no verb).

19. Karlsson 2007.

20. Evans 2003, 633.

21. See especially Evans and Levinson 2009 and Evans 2009.

22. These various figures are from Evans 2009.

23. One can see the same processes at work in modern English, although they are no doubt held in check by the unifying influences of the media—radio, television, newspapers, the Internet. Young people seek new expressions to differentiate themselves from their elders, gangs develop gangspeak, and racial groups evolve ways of speaking that cannot be attributed wholly to a different first language. African American English, for example, cannot be attributed to an indigenous African language. Women develop ways of speaking that differ from the more gruff intonations of malespeak.

24. Evans 2009, 46.

25. Pinker and Bloom 1990, 715.

26. Tomasello 2003, 5.

27. Karlsson 2007.

28. Blatt 1957.

29. Chomsky 2010, 60.

30. The term *grammaticalization* is often taken to refer to processes that arise in languages that already have grammar. This implies a jump from protolanguage to language, which to my mind contains vestiges of "big bang" theory. An alternative view, which I prefer, is that language—and grammar—emerged gradually, so the very concept of protolanguage is unfounded. In NSL and ABSL, we are effectively witnessing the emergence of new languages. For an alternative viewpoint, though, see Arbib 2009.

31. For more detailed exposition, see Hopper and Traugott 2003, and Heine and Kuteva 2007.

32. Senghas, Kita, and Özyürek 2004.

33. See Kirby and Hurford 2002 for a review.

34. Stokoe, Casterline, and Croneberg 1965.

35. Aronoff et al. 2008.

36. Hockett 1960.

37. Hockett 1960, 90.

38. Aronoff 2007. Aronoff is proud to call himself an "antidecompositionalist"—one who is opposed to the view that words can be broken down into parts. Of course words *can* be broken down into parts, but Aronoff argues that this is not the way people actually use words, or think about them. The issue is complex and somewhat technical, and readers are referred to Aronoff's article for more

extensive discussion. Interestingly, Aronoff appeals to the "lexicalist" position described by Chomsky (1970) as support for his view.

39. Of course it's more complicated than that, since English has evolved in many ways and includes lots of borrowing. The suffix *-ed* derives from ancestral Proto-German.

40. Givón 1971, 413.

41. Pinker 1994.

42. This phrase was coined by Marler (1991) to account for the acquisition of birdsong, but borrowed by Locke and Bogin (2006) to account for the human acquisition of language.

43. Everett 2005, 622.

44. Christiansen and Chater 2008.

45. Hauser, Chomsky, and Fitch 2002.

46. See Jackendoff 2002, 204.

CHAPTER 3

Do Animals Have Language?

1. Saki 1936, 122; first published in the collection *The Chronicles of Clovis* in 1911.

2. Jürgens 2002.

3. Provine 2000.

4. Goodall 1986, 125.

5. Arcadi, Robert, and Boesch 1998.

6. Schaller 1963.

7. Slocombe and Zuberbühler 2007.

8. Hopkins, Taglialatela, and Leavens 2007.

9. It is commonly believed that chimpanzees are a good model for the common ancestor we share with them. As we shall see in chapter 11, this notion is changing. Chimpanzees have probably changed as much as humans have since the split some six or seven million years ago.

10. Snowdon 2004, 132.

11. Although, confusingly, it has also been maintained that gorillas have the most complex and frequent vocalizations of the great apes (Harcourt and Stewart 2007).

12. Oddly enough, there is a novel by the late New Zealand writer Nigel Cox, called *Tarzan Presley*, in which Elvis is raised by gorillas in the Wairarapa district of New Zealand, before moving to the United States and establishing his career as a singer and rock star. As far as I know, there are no gorillas outside of zoos in New Zealand, although one or two may have invaded the rugby scrum.

13. Aitken 1981; Sutton, Larson, and Lindeman 1974.

14. MacLean and Newman 1988.

15. Hihara et al. 2003.

16. Roy and Arbib 2005.

17. Jarvis 2006.

18. Knight 1998.

19. Dawkins and Krebs 1978.

20. Or else, heaven help us, the sound of screwtops being unscrewed, since corks seem to be disappearing from Australian (and New Zealand) wine bottles. On the whole, I think this is a positive development, since I have yet to hear patrons complaining to a waiter that the wine is screwed.

21. Cheney and Seyfarth 1990.

22. This does seem a little difficult to square with Jarvis's idea that learned calls are more likely than fixed calls to attract predators.

23. In this case it *was* the rock star who got the girl.

24. Mithen 2005. I am not entirely convinced. With very few exceptions, people can talk, but very few of us can sing.

25. Deacon 1997, 225.

26. Blowing his own trumpet, so to speak; see http://www.reuters.com/news/video/videoStory?videoId=1231.

27. Kenneally 2007.

28. Pepperberg 2000.

29. This may be a little unfair. Sheldrake and Morgana (2003) have provided a more detailed account in a journal called *Journal of Scientific Exploration.* You be the judge.

30. Chimps and bonobos shared a common ancestry with humans until six or seven million years ago. Chimps and bonobos split into separate species about two million years ago.

31. Arcadi (2000), who may have overlooked the vocal exchanges between politicians in his analysis.

32. Hayes 1952.

33. Ladygina-Kohts was perhaps the true pioneer of the study of chimpanzee language. She began her work in Moscow in 1913, and continued through the Russian Revolution, publishing her major work in Russian in 1935. The English translation did not appear until 2002.

34. This may seem to contradict the idea that chimp vocalizations can't be learned. However, chimpanzees do emit a variety of bark sounds, including a pant hoot bark and a waa-bark, so Ladygina-Kohts may have mistaken a natural chimpanzee sound for an imitation.

35. Ladygina-Kohts 2002, 380. It should be said, though, that, Jodi was disadvantaged in ways other than being a chimpanzee. He was emotionally and physically deprived, despite Ladygina-Kohts's attentions, and died at a young age. He was also a prerevolutionary child, whereas Roodi was born well after the Russian Revolution.

36. The often striking intelligence of dogs may have to do with a long history of domestication by humans.

37. Kaminsky, Call, and Fischer 2004.

38. Savage-Rumbaugh, Shanker, and Taylor 1998.

39. I observe it in myself, too, when in Italy or France, where my level of understanding is not too bad, but I struggle to find the words to make my own contribution to the conversation.

40. If you repeat the word *rest*, without pausing between repetitions, you may find that it has transmuted into the word *stress*. It may even transmute further into repetitions of the word *ester*.

41. Human infants, though, can learn to segregate words at an early age. By about one year of age, they can distinguish function words (like *a, the, that*, etc.) from content words (Shi, Werker, and Cutler 2006).

42. I am told that this term may now be politically incorrect, and one should call it "parentese."

43. It might be interesting to insert these words in an otherwise meaningless sentence to see how Kanzi responds.

44. Neidle et al. 2000.

45. In his novel *The Thirteen-Gun Salute*, Patrick O'Brian assumes the animal was an orangutan.

46. Gardner and Gardner 1969.

47. Patterson 1978.

48. This sounds enigmatic, but "child-side" is the local name for the place in the laboratory where children were studied.

49. Pinker 1994, 340.

50. Kanzi has since coauthored an article, along with two of his relatives, Panbanisha and Nyota (Savage-Rumbaugh et al. 2007). The corresponding author, though, is Sue Savage-Rumbaugh.

51. Miles 1990.

52. Herman, Richards, and Wolz 1984.

53. Pepperberg 1990.

54. Bickerton 1995.

55. Premack 1988.

56. Jackendoff 2002. The implication behind protolanguage is that language emerged in a single step from that point, consistent with "big bang" theories of language evolution. Evidence on grammaticalization, discussed in the previous chapter, suggests that grammatical language evolved gradually. To my mind, this casts doubt on the notion of protolanguage.

57. Köhler 1925.

58. To argue that communicative gestures are a form of problem solving is not to deny them a role in the evolution of language. Problem solving may be indeed have sown the seeds for the later emergence of syntax.

59. Tomasello 1999.

60. Povinelli 2001.

61. Whiten, Horner, and de Waal 2005.

62. Tanner and Byrne 1996.

63. Tomasello et al. 1997.

64. Pollick and de Waal 2007.

65. For gorillas see Pika, Liebal, and Tomasello 2003, chimpanzees see Liebal, Call, and Tomasello 2004, and bonobos see Pika, Liebal, and Tomasello 2005. See also Arbib, Liebal, and Pika 2008 for a discussion of these matters, especially in relation to the origins of language.

66. Chomsky 1966, 78.

67. Hauser, Chomsky, and Fitch 2002.

68. Terminology can be confusing. As far as I can tell, *FLN*, *I-language*, and *universal grammar* all refer to essentially the same thing. The distinction between FLB and FLN seems a remarkably clumsy way of saying that human and animal language overlap in some respects but not in others.

69. Hauser, Chomsky, and Fitch 2002, 1571.

70. Why starlings? Perhaps the authors of this study were inspired by *Henry IV*, where Shakespeare has Hotspur say

> Nay,
> I'll have a starling be taught to speak
> Nothing but "Mortimer," and give it him
> To keep his anger still in motion.

71. Gentner et al. 2006. This article was published in the prestigious journal *Nature*, and soon attracted attention in the media. The experiment was actually based on an earlier one by Fitch and Hauser (2004), in which tamarins proved unable to discriminate the embedded sequences. These authors made no claim that the test had anything to do with recursion as such, but rather compared the animals' ability to distinguish a finite-state grammar (repeated pairs) from a phrase-structure grammar (embedded pairs), but they appear to accept that the phrase-structure grammar could be parsed by comparing strings of successive elements. To my mind, this somewhat trivializes their study, although it does make a technical point.

72. See, for example, Thompson 1969.

73. I have explained all this more fully in Corballis 2007b. The original article on tamarins was published in *Science*, and the later article on starlings by *Nature*, but neither of these eminent journals would accept a critical commentary. I regard the technique used by the authors as a scientific virus that needs to be eradicated. The technique has also been picked up by researchers using brain imaging to find the spot in the brain responsible for recursive parsing (Bahlmann, Gunter, and Friederici 2006; Friederici et al. 2006), but needless to say such endeavors are doomed to likely failure, and your intrepid author has been quick to pounce on this as well (Corballis 2007a).

CHAPTER 4

How Language Evolved from Hand to Mouth

1. Chomsky 2010, 58.

2. Bickerton 1995, 69. He has since modified his view somewhat, arguing that the roots of syntax might be traced to reciprocal altruism in primates, but he still appears to maintain that language in the genus *Homo* was essentially protolanguage, without syntax, until the emergence of *Homo sapiens* (Calvin and Bickerton 2000). More recently still, Bickerton (2010) outlines a scenario whereby language evolved more gradually during the Pleistocene.

3. Klein 2008, 271. The nature and implications of the archeological record are discussed more fully in chapter 12.

4. Crow (2010) goes so far as to locate the genetic basis of human speciation in a particular gene pair, *Protocadherin11XY*, located in homologous regions of the X and Y chromosomes. According to this account, Prometheus was indeed a male, since the critical event occurred on the Y chromosome. Crow suggests that this event occurred within the past 200,000 to 150,000 years, and was the basis of human speciation.

5. Pinker and Bloom 1990, 708.

6. Pinker and Bloom 1990, 711.

7. Ploog (2002) shows how the neural basis for vocalization in humans differs from that in nonhuman primates.

8. Condillac 1971.

9. Condillac 1971, 172.

10. Condillac 1971, 174.

11. Condillac 1971, 175–176.

12. Darwin 1896, 87; emphasis added.

13. Wundt 1900.

14. Critchley 1975, 221.

15. Klima and Bellugi 1979; Poizner, Klima, and Bellugi 1987. Some are still resistant to the idea that signed languages are true languages. At a recent conference I presented the gestural theory, only to be told by a prominent linguist that signed languages were merely pantomime.

16. Two more recent books continue the trend. One is Armstrong and Wilcox's (2007) book *The Gestural Origin of Language*, and the other Rizzolatti and Sinigaglia's (2006) *Mirrors in the Brain*, which is built on the discovery of mirror neurons.

17. Ramachandran 2000.

18. Arbib and Rizzolatti 1997; Rizzolatti and Arbib 1998.

19. Binkofski and Buccino 2004.

20. As an aside, so to speak, it is worth mentioning here that the system in humans is generally biased to the left side of the brain. For some speculation about this, see Corballis 2004a.

21. Rizzolatti, Fogassi, and Gallese 2001.

22. Dick et al. 2001.

23. Arbib 2005; 2010. Grodzinsky (2006) has expressed reservations about the role of mirror neurons in language.

24. Liberman et al. 1967. It should be noted, though, that the motor theory of speech perception is still controversial after more than 40 years, as is the role of mirror neurons—see Hickok 2009 and Lotto, Hickok, and Holt 2009.

25. The perception of phonemes as invariant despite acoustic variation may depend on a region in the left inferior frontal sulcus—an area not far from Broca's area (Myers et al. 2009).

26. Kohler et al. 2002.

27. Rizzolatti and Sinigaglia 2006.

28. Fadiga et al. 1995.

29. Aziz-Zadeh et al. 2006.

30. Xua et al. 2009.

31. Pettito et al. 2000.

32. Pietrandrea 2002.

33. Emmorey 2002. Although some signs are iconic, signers nevertheless distinguish signs from free pantomime, and aphasia for sign language leaves pantomime unaffected (e.g., Marshall et al. 2004). Signers don't seem to notice that some signs are iconic and some are not.

34. Pizzuto and Volterra 2000.

35. Burling 1999.

36. Saussure 1977.

37. Pinker 2007.

38. Shintel, Nusbaum, and Okrent 2006. See also Shintel and Nusbaum 2007 for evidence that people respond to pictures more quickly if a spoken sentence describing the picture matches the motion represented in the picture. They match a moving object, such as a galloping horse, more quickly to the sentence if the sentence is spoken quickly, and match a stationary object more quickly if the sentence is spoken relatively slowly.

39. Hockett 1978, 274–275.

40. Frishberg 1975.

41. In Plato's *Cratylus*, Socrates asks, "Suppose we had no voice or tongue, and wanted to communicate with one another, should we not, like the deaf and dumb, make signs with the hands and head and rest of the body?"

42. Zeshan 2002.

43. Evans 2009.

44. Pinker 2007.

45. Wood and Collard 1999.

46. Burling 2005, 123.

47. MacNeilage 2008. He is a compatriot of mine, but I'm sure I can bring him round.

48. Rizzolatti et al. 1988.

49. Petrides, Caddoret, and Mackey 2005.

50. Gentilucci et al. 2001. You are advised to try to keep your mouth shut when reaching for a large object, such as a pineapple.

51. See Gentilucci and Corballis 1996 for a review.

52. Bernardis et al. 2008.

53. McGurk and MacDonald 1976. At the time of writing, and I hope still, you can experience the McGurk effect on http://www.youtube.com/watch?v=aFPtc8BVdJk.

54. Calvert and Campbell 2003; Watkins, Strafella, and Paus 2003.

55. The edited volume by Sutton-Spence and Boyes-Braem (2001) spells out systems used in several European signed languages.

56. Emmorey 2002.

57. Muir and Richardson 2005.

58. Studdert-Kennedy 1998.

59. Browman and Goldstein 1995.

60. Vargha-Khadem et al. 1995.

61. Fisher et al. 1998; Lai et al. 2001.

62. Corballis 2004a.

63. Liégeois et al. 2003.
64. Haesler et al. 2007.
65. Groszer et al. 2008.
66. Enard et al. 2009. The vast congregation of authors suggests that the effect may be due to alterations in circuits linking the cortex to the basal ganglia. This may have some bearing on the manner in which vocalization was brought under cortical control in humans.
67. Enard et al. 2002.
68. Now deceased.
69. Krause et al. 2007.
70. Coop et al. 2008.
71. Evans et al. 2006.
72. Some of them evidently had red hair and fair complexions (Culotta 2007). Make of that what you will. Recent evidence from sequencing the Neandertal genome strongly suggests that the ancestors of non-Africans probably did mate with Neandertals, so there was some limited exchange of DNA (Green et al. 2010).
73. Schroeder and Myers 2008.
74. See Fisher and Scharff 2009 for a recent review.
75. P. Lieberman 1998; Lieberman, Crelin, and Klatt 1972.
76. Noonan et al. 2006.
77. See, for example, Boë et al. (2002; 2007), who argue, contrary to Lieberman, that the vocal apparatus of the Neandertals would indeed have permitted articulate speech, although his arguments have been questioned in turn by de Boer and Fitch (2010). Boë et al. (2007) also suggest that any restrictions on articulation, in human infants as in Neandertals, may have been due to imprecise motor control of the articulators rather than to the shape of the vocal tract.
78. Tattersall 2002, 167. See also previous note.
79. D. E. Lieberman 1998.
80. Lieberman, McBratney, and Krovitz 2002.
81. In other words, our foreheads are kind of bulbous.
82. Kay, Cartmill, and Barlow 1998.
83. DeGusta, Gilbert, and Turner 1999.
84. MacLarnon and Hewitt 2004.
85. P. Lieberman 2007, 39. Robert McCarthy of Florida Atlantic University has recently simulated how the Neandertal would have sounded when articulating the syllable /i/ (or ee), based on the shape of the vocal tract. It can be found on http://anthropology.net/2008/04/16/reconstructing-neandertal-vocalizations/, and compared with a human articulating the same sound. One observer described the Neandertal's attempt as sounding more like a sheep or a goat than a human. But see Boë et al. 2002 for contrary evidence.
86. Konner 1982.
87. Kingsley 1965, 504.
88. The eloquent Italians, for example, gesture more than we undemonstrative Kiwis do.
89. Russell, Cerny, and Stathopoulos 1998.
90. Evans 2009.

91. An even more parsimonious system is Silbo Gomero, a whistled language used by shepherds on the island of Gomero in the Canary Islands, which is reduced to two vowels and four consonants. It is perhaps an unfair example, though, because it is essentially a cut-down version of Spanish (Carreiras et al. 2005).

92. Everett 2005.

93. Salmond 1975, 50.

94. It depends on how the analysis is done (Evans 2009).

95. Saying it backwards probably helps a bit.

96. The commonly expressed idea that infants can discriminate phonemes of *all* the world's languages ("Universal Theory") has been challenged by Nittrouer (2001), but defended by Aslin, Werker, and Morgan (2002).

97. Darwin 1896, 89.

98. The cooking of food, not televisions, although given the explosive language of some TV chefs one might wonder.

99. Corballis 2002.

100. McNeill 1992; Goldin-Meadow and McNeill 1999.

101. Wittgenstein 2005.

PART 2

Mental Time Travel

1. Hockett 1960.

CHAPTER 5

Reliving the Past

1. Forster 1995, 133–134.

2. Bruce, Dolan, and Phillips-Grant 2000.

3. The distinction between episodic and semantic memory was developed by the Canadian psychologist Endel Tulving (1983; 2002).

4. That is as I remember it, of course. Others in the class may remember it differently.

5. See, for example, Tulving et al. 1988.

6. Tulving 2002.

7. Burianova and Grady (2007) examined brain activation, using functional magnetic brain imaging (fMRI), while people retrieved autobiographical, episodic, and semantic memories. There was considerable overlap, suggesting common processes, but each kind of memory also elicited unique activation. Autobiographical retrieval triggered unique activation in the medial frontal lobe, which is probably associated with representation of the self. Episodic retrieval uniquely activated the right middle frontal lobe and semantic retrieval the right inferior temporal lobe.

8. Wearing 2005.

9. Short-term memory, also known as working memory, holds information in consciousness for a few seconds, and is distinct from the semantic and episodic systems that make up long-term memory.

10. This case was first described by Scoville and Milner (1957). For a more recent account, see Corkin 2002.

11. This theory was developed by Larry Squire and colleagues (e.g., Squire 1992), but other models of hippocampal function have been suggested (e.g., Moscovitch et al. 2006).

12. See Tulving 2002 for a review.

13. Tulving 2001.

14. Hodges and Graham 2001.

15. Mitchell 2006.

16. Loftus and Loftus 1980.

17. But not necessarily sober.

18. Kundera 2002, 122–123.

19. Treffert and Christensen 2006.

20. Luria 1968, 22.

21. Parker, Cahill, and McGaugh 2006.

22. Pinker 2007 bases this estimate on the number of words in a desk (i.e., not an unabridged) dictionary.

23. Loftus and Ketcham 1994, 39.

24. Roediger and McDermott 1995.

25. Burnham 1989.

26. Bernheim 1989, 164–165.

27. In later editions of the book (e.g., 1994), Bass and Davis qualified their statement, indicating that symptoms of distress need not imply abuse. The hysteria over sexual abuse has largely abated since the 1990s, although there are undoubtedly still many innocent individuals whose lives were wrecked because of false accusations arising from falsely recovered memories.

28. The question of whether a memory is true or false is a classic one in a branch of science known as *signal detection theory*. Memory is often a weak signal, and one can be as uncertain of whether a memory is real or not as of whether a noise in the house is an intruder or not, or whether a pain in one's chest is a signal of an impending heart attack. Where a signal is weak, there are two kinds of error one can make: one can fail to detect a signal that is truly there, or one can falsely detect a signal that is not there. If we apply that understanding to the question of memory for sexual abuse, one can suppose that abuse did not occur when in fact it did, or one can suppose that it did occur when it did not. Each error carries a cost. Failure to detect past abuse may leave a perpetrator free to continue to abuse, while false detection can result in innocent people being punished. Much of the political argument boils down to the question of which of these errors is the more costly. Many feminists appear to believe that it is better that some innocent people be sent to jail than to have perpetrators of abuse at liberty, while the basic legal tenet that one is innocent until proven guilty protects the innocent at the expense of failing to detect actual perpetrators. Judgments may be biased toward the former by aggressive therapy leading to implantation of false memories.

29. Hood 2001.

30. Pinker 1994.

31. The story of Genie and other so-called wild children is told in Newton 2004.

32. Pavlov 1927. Classical conditioning featured as a major theme in Aldous Huxley's *Brave New World*.

33. In the politically correct terminology of modern experimental psychology, the term "participant" is preferred to "subject." In this particular case, the term "subject" seems altogether more appropriate.

34. Watson and Rayner 1920.

35. Skinner 1957.

36. Skinner 1962. The title derives from Thoreau's *Walden*.

37. It was a word I had only recently discovered, and I rather liked the sound of it. I really didn't know what it meant.

CHAPTER 6

About Time

1. Suddendorf and Corballis 1997; 2007.

2. Atance and O'Neill 2005. Suddendorf (2010) also suggests the term "episodic foresight."

3. See also Busby Grant and Suddendorf 2009.

4. Atance and O'Neill 2005; Klein, Loftus, and Kihlstrom 2002.

5. Ingvar 1979, 21.

6. Schacter, Addis, and Buckner 2007.

7. Addis, Wong, and Schacter 2007; Okuda et al. 2003; Szpunar, Watson, and McDermott 2007.

8. Botzung, Denkova, and Manning 2008; D'Argembeau et al. 2008; Hassabis, Kumaran, and Maguire 2007.

9. The constructive nature of episodic memory was classically demonstrated by the British psychologist Sir Frederic C. Bartlett (1932).

10. Suddendorf and Corballis 1997; 2007.

11. Köhler 1925.

12. See, for example, Kamil and Balda 1985.

13. This inspired term was proposed by Suddendorf and Busby (2003).

14. Clayton, Bussey, and Dickinson 2003.

15. Dally, Emery, and Clayton 2006.

16. Ferkin et al. 2008.

17. Roberts et al. 2008.

18. Bischof 1978; Bischof-Köhler 1985; Suddendorf and Corballis 1997.

19. Correia, Dickinson, and Clayton 2007.

20. See Suddendorf, Corballis, and Collier-Baker 2009 for a critique of this and other studies purporting to refute the Bischof-Köhler hypothesis.

21. McGrew 2010.

22. Hunt and Gray 2003.

23. This is the consensus of Whiten and eight coauthors (1999). See also Whiten, Horner, and de Waal 2005.

24. Boesch and Boesch 1990.

25. Mulcahy and Call 2006. Curiously, though, bonobos, like gorillas, show little evidence of tool use in the wild (McGrew 2010).

26. See Suddendorf 2006 for a critique.

27. Bang on.

28. It has been suggested that the linear concept of time did not emerge until late antiquity, and that earlier understanding was cyclical (Butterfield 1981).

29. Everett 2005.

30. Pettit 2002.

31. Andrews and Stringer 1993.

32. Markus and Nurius 1986.

33. James 1910.

34. Markus and Nurius 1986, 954.

35. Neisser 2008, 88.

CHAPTER 7

The Grammar of Time

1. From a conversation with Freddy Gray, reported in *The Spectator* of 10 April 2010.

2. Pinker 2003, 27.

3. Yes, I know—even this book is something of a just-so story.

4. Hebb had a distinctly mediocre undergraduate record, and his initial intention was to become a novelist.

5. See also Corballis and Suddendorf 2007.

6. Tulving 2002.

7. Pinker 2007. As I mentioned before, this estimate is based on the number of words in a common dictionary.

8. Deacon 1997.

9. Liszkowski et al. 2009.

10. Lin 2006. Aspectual markers are formally distinguished from tense, and have to do with the temporal flow rather than the location in time. In English, for example, the sentences "I talk" and "I am talking" are both present tense, but are distinguished by aspect, with the first representing a habitual activity and the second a progressive or continuous one.

11. It's time I got this damn book written.

12. Reichenbach 1947.

13. Núñez and Sweetser 2006.

14. Chen 2007.

15. Everett 2005. It should be said that Everett's analysis is controversial, as evident in the commentaries following his article. Yet he and his family are the only outsiders who know the language of the Pirahã, and are best equipped to pronounce on its properties, at least to those who don't speak it.

16. In some respects Everett is not entirely consistent. For instance he does note that the Pirahã are "afraid of evil spirits" (2005, 623). To the Western eye, evil spirits might well be seen as fictitious, although perhaps to the Pirahã they are seen as part of everyday reality. Again, Everett states that "Pirahã repeat and embellish these stories" (633), but the claim is that the stories are based on firsthand experience, and are not fictional.

17. Everett 2005, 632.

18. Whorf 1956, 57–58.

19. Malotki 1983. It may well be that Everett also underestimated the Pirahã sense of time.

20. They should be so lucky.

21. Skinner 1957.

22. Westen 1997, 530. Nevertheless Skinner was interested in psychoanalysis, and even wanted to be psychoanalyzed—but was turned down! (Overskeid 2007).

23. This was the main theme of Chomsky's famous 1959 review of Skinner's 1957 book *Verbal Behavior*.

24. Frege 1980, 79. To most linguists, though, words are but a step in the hierarchy that proceeds from phonemes to morphemes to words. As we saw in chapter 2, though, words may be the true primitives in an evolutionary sense, with phonology and morphology emerging as a result of pressure to create distinctions (Aronoff 2007).

25. Horne Tooke 1857. Objects are represented by nouns and actions by verbs. Technically, though, the object/action distinction is not the same as the noun/verb distinction. Many nouns (such as *love* or *coherence*) do not represent objects, and many verbs (such as *enjoy* or *wonder*) do not represent actions. The idea that the first words were nouns and verbs actually goes back to Plato.

26. In case you're traveling this year, they are Warao in Venezuela, Nadëb in Brazil, Wik Ngathana in northeastern Australia, and Tobati in West Papua New Guinea.

27. Aronoff et al. 2008.

28. A curious exception is C. P. Snow's last novel, *A Coat of Varnish*, published in 1979. The murderer is never revealed.

29. It can go the other way. The 1994 movie *Heavenly Creatures*, directed by Peter Jackson, is based on the true story of two New Zealand schoolgirls who murdered the mother of one of them. They were of course caught, and one of them is now an internationally known author of murder stories.

30. Maybe we should add quarks and leptons and bosons, and the other entities postulated by modern theoretical physics.

31. Wilson 2002, 64.

32. Boyd 2009.

33. There must surely be limits to this. Science depends on the discovery of truth and, one may hope, is adaptive. This adaptiveness may be foiled by belief in falsehood.

34. Or so I believe, God help me.

35. Boyd 2009, 206.

CHAPTER 8

Mind Reading

1. Randi 1982. The James Randi Educational Foundation was established in 1996 to further Randi's work. It offers a large monetary prize to anyone who can

demonstrate psychic powers. As of 10 July 2007 this long-standing offer remains unclaimed, and the award stands at $1,000,000. See www.randi.org for updates.

2. Marks and Kammann 1980.

3. Sheldrake 1999. Once again, the intrepid David Marks has struck, in the new edition of his earlier book with Kammann (Marks 2000). One wonders if Sheldrake saw it coming.

4. Darwin 1872, 357—a long quote, to be sure, but the image of Darwin pulling faces and trying to look savage is too much to resist.

5. Piaget 1928.

6. Borke 1975. One might have thought, though, that mountains would have been pretty familiar to Swiss children.

7. Southgate, Senju, and Csibra 2007. The authors included a familiarization phase to ensure that the infants would look to where the actor would retrieve a ball when it was actually present, as well as variations to rule out other possibilities, such as the infants simply looking to where the ball was most recently placed.

8. I suppose the catwalk might make you wonder, although those young women seem scarcely built for reproduction.

9. Cosmides and Tooby 1992.

10. Trivers 1974.

11. Barkow, Cosmides, and Tooby 1992.

12. The term was inspired by Franz B. M. de Waal, whose 1982 book *Chimpanzee Politics* noted that some of the social strategies used by chimpanzees had a Machiavellian flavor. Much has been written on the question of whether chimpanzees and other primates are truly Machiavellian, or possess what has been termed "theory of mind"—the ability to take on the mental perspectives of others (e.g., Byrne and Whiten 1990; Premack and Woodruff 1978; Tomasello and Call 1997; Whiten and Byrne 1988). Whatever the case with respect to other primates, it seems that we humans are supreme in our ability to lie, cheat, and deceive, while also maintaining outward respectability.

13. Dennett 1983.

14. Cargile 1970.

15. Dunbar 2004, 185.

16. Fifth-order recursion may be necessary for religious belief, but is surely not sufficient. I think I am capable of this order of recursion, but I am not religious. I can't answer for Robin Dunbar.

17. Baron-Cohen 1995.

18. Grandin 1996; Grandin and Barron 2005; Grandin and Scariano 1986. Grandin would now be classified as a case of Asperger's syndrome, which is a form of autism in which intellectual function is high.

19. Sacks 1995.

20. Grandin and Johnson 2005.

21. Senju et al. 2009.

22. Crespi and Badcock 2008.

23. Actually best known for his involvement in the antipsychiatry movement.

24. Hamilton 2005, 205.

25. Baron-Cohen 2002.

26. See Crespi and Badcock 2008 for a more detailed account.

27. See Crespi and Badcock 2008, 248, for more extensive lists. This account, though, is not uncontroversial, since some psychiatrists regard schizophrenia and autism as related, rather than as polar opposites. Both show very similar patterns of brain activation (Pinkham et al. 2008). One possibility is that the negative symptoms of schizophrenia join autism at one end of the spectrum, with positive symptoms of schizophrenia at the other (van Rijn, Swaab, and Aleman 2008). Recent evidence implicates genetic influences in both autism (no fewer than three papers on this appear in the 28 May 2009 issue of *Nature*) and schizophrenia (e.g., Esslinger et al. 2009), with no indication that imprinting plays a role. Even so, Crespi and Badcock present an interesting scenario with implications beyond autism and schizophrenia.

28. Badcock and Crespi 2006.

29. Baron-Cohen 2009.

30. The reader should appreciate that the distinction between people people and things people is bound to be somewhat simplistic.

31. Maudsley 1873, 64.

32. Kéri 2009. The gene in question is *Neuregulin 1*, and the particular genotype is T/T.

33. Horrobin 2003. See Richmond 2003 for a perspective on David Horrobin.

34. Farmelo 2009. In Bristol, UK, Dirac's reputation is overshadowed by that of his schoolmate Archie Leach, better known later as Cary Grant.

35. Langford et al. 2006.

36. Wechkin, Masserman, and Terris 1964.

37. de Waal 2008.

38. See Povinelli, Bering, and Giambrone 2000 for a summary.

39. Tomasello, Hare, and Agnetta 1999.

40. Povinelli and Bering 2002.

41. Hare et al. 2000. Marmosets also choose food that a watching dominant marmoset can't see (Burkart & Heschl, 2007).

42. Hare, Call, and Tomasello 2001.

43. Hare, Call, and Tomasello 2006.

44. Hare and Tomasello 1999.

45. The cleverness of dogs has become a matter of some contention. The apparent ability of domestic dogs to read human intentions has been attributed to selective breeding, and may be rather more restrictive than is at first apparent. Dogs have been bred to collaborate with people, but they don't seem to collaborate with other dogs. For a useful discussion, see Morell 2009.

46. For a detailed critique of Hare's studies, and of others studies claiming theory of mind in corvids, see Penn, Holyoak, and Povinelli 2008.

47. Whiten and Byrne 1988.

48. Leslie 1994; Tomasello and Rakoczy 2003.

49. Leslie (1994) has referred to these levels of attribution as ToMM-1 (Theory of Mind Module 1), involving attribution of goal-directedness and self-generated motion, and ToMM-2 (Theory of Mind Module 2), involving full-blown theory of mind. Hauser and Carey (1998) remark that "the intellectual tie breaker between humans and nonhumans is likely to lie within the power of ToMM-2."

50. Gallup 1998.

51. For review, see Suddendorf and Collier-Baker 2009.

52. Penn, Holyoak, and Povinelli 2008, 129.

53. No doubt having in mind, as it were, the onetime preeminence of the Rolls-Royce automobile manufacturing company.

CHAPTER 9

Language and Mind

1. Fodor 1975.

2. Pinker 2007, 90.

3. For the benefit of U.S. readers, "pissed" here refers to being drunk, and not to being angry. It might be appropriate for newts to describe the occasionally drunken newt as being as pissed as a human.

4. This act has created considerable controversy and argument as to how it should be interpreted. It does seem to permit hunting by women.

5. Watson 1913, 158.

6. Griffin 2001, 1.

7. Inoue and Matsuzawa 2007. The chimps first learned to recognize the digits 1 through 9, and to point to them in succession when they were randomly displayed on a console. In one memory test, five digits were randomly selected and displayed in random positions and then blanked out and replaced with white squares. One chimp, Ayumu, maintained 80 percent accuracy when the duration of the digits was reduced to a mere fifth of a second, a performance well above that reached by a group of university students.

8. Pinker 2007, 23.

9. Fauconnier 2003, 540.

10. Grice 1989, 30–31. If you're not used to minding your philosophical Ps and Qs, you could take as an example P = "Hello, John, this *is* a surprise. How were they?" and Q = This is my friend John, who must be back early from Australia, where he went to visit his aging parents.

11. Sperber and Wilson 2002, 15.

12. Sperber and Wilson 1986.

13. The term "minimalism" is more often used with respect to a school of music, in which the work is stripped down to its basic features.

14. Grice 1975.

15. Language deficits can be identified in infants as young as two, and are manifest in poor imitation and gestural communication (Luyster et al. 2008).

16. Irony may be culturally specific, despite Kierkegaard's claim. Tom Suddendorf, a native of Germany, tells me that Germans don't need it—and even if they did, they wouldn't get it. He may have been being ironic.

17. Perhaps the only example of a double positive translating as a negative.

18. Dostoevsky 2008.

19. Gibbs 2000.

20. Papp 2006. GCSE stands for General Certificate of Secondary Education, a qualification sought by 12- to 14-year-old children in England, Wales, and

Northern Ireland (but not Scotland). It can be taken in several subjects, but only once.

21. Happé 1995.

22. As an addict of cryptic crosswords, I often need to assume the somewhat autistic mentality of the compiler. One normally assumes the word *number*, for example, has to do with counting, but in crosswords it often refers to anesthetic. And don't be fooled by the word *flower*, which can be synonymous with *river*. Or *layer*, which can refer to a hen.

23. Walenski et al. 2008.

24. Hauser, Chomsky, and Fitch 2002.

25. Tomasello 2008.

26. One wonders, though, why they can point at all, since for most of their evolutionary existence they have not had to put up with the company of humans.

27. Tomasello 2008, 55.

28. Rivas 2005.

29. Greenfield and Savage-Rumbaugh 1990.

30. I am sharing this information with you, and don't expect any reward. It would be nice, though, if you would consider buying the book, if you haven't done so already.

31. The sex of the child is not stated, but I am assuming she is female. My assumption is based on the fact that I have a granddaughter. This is also information I'd like to share with you. *Note added in 2010:* I now have three granddaughters.

32. But how would you know?

33. de Villiers (2009) provides a useful discussion of the joint emergence of language and theory of mind, and the interface between them.

Chapter 10

The Recurring Question

1. From *Pensées* 1670.

2. Walsh et al. 2003.

3. Johnson 2000.

4. Australia is a great slab of a place to the west and somewhat to the north of New Zealand.

5. Sosis 2004.

6. Bloom (2004) is not the only one to suggest that dualism is innate. See also Shermer's *Why People Believe Weird Things* (1997), and Hood's *Why We Believe the Unbelievable* (2009).

7. Dennett 1995.

8. Dobzhansky 1973, 125.

9. See "Shunning the E-word in Georgia," *Science*, **303**, 759 (2004).

10. This is not to say that all religions support the idea. Writing in the 18 January 2005 issue of *L'Osservatore Romano*, the official Vatican paper, Fiorenzo Facchini argued that intelligent design belongs to the realms of philosophy and religion, but not of science. He writes that "it is not correct from a methodological point of view to stray away from the field of science while pretending to do

science." The Vatican has for many years tolerated the teaching of evolutionary theories, and in 1950 a papal encyclical officially permitted Catholics to discuss Darwin's theory of evolution. In another development, on 20 December 2005 federal district court judge John Jones III ordered the schools in Dover, Pennsylvania, to remove references to intelligent design from the science curriculum, on the grounds that it is not science.

11. As reported in *Time* magazine, 15 August 2005, 47.

12. Davis and Kenyon 2004, 99–100.

13. I am told the octopus eye is better designed, with the optic nerve located at the back of the retina.

14. Not everyone believes this, but few of the dissenters are willing to attribute the plays to an ape. The poems, maybe, but not the plays.

15. Cited by Darwin 1896, 49.

16. Quoted in Marchant 1916, 241.

17. Gross 1993.

18. A word that itself gives the game away.

19. And for insects, of course. To put it another way, they took hold of an idea and flew with it.

20. Chomsky 1966.

Chapter 11

Becoming Human

1. Until fairly recently, the term *hominid* was used, but the great apes were invited to join the hominid clade when it was discovered just how closely their genetic makeup resembled our own. See also Note 33 in chapter 2.

2. That is, if we discount the role of horses, mules, carriages, bicycles, scooters, automobiles, trains, ships, airplanes, and space rockets. Oh, and swimming, I suppose—but you get what I mean.

3. Brunet et al. 2002.

4. Sibley and Ahlquist 1984. There is nevertheless still some uncertainty and conflicting evidence about the ape-hominin split. One recent theory, based on analysis of diversity of differences across the genome, is that the hominin and chimpanzee lineages split close to seven million years ago, but then hybridized, finally splitting again within the past 6.3 million years (Patterson et al. 2006).

5. Galik et al. 2004.

6. Leakey 1979.

7. Thorpe, Holder, and Crompton 2007.

8. Lovejoy et al. 2009.

9. Quoted in Gibbons 2009, 39.

10. These developments may help destroy the idea that the chimp somehow failed to move on from common ancestry with humans some six million years ago, and is therefore a sort of failed human. Chimpanzee evolution simply took a different path.

11. Teleki 1973.

12. Bipedalism is a pain in the ass.

13. Formerly known as Ayer's Rock.

14. I am indebted to the useful discussion in chapter 12 of Michael Sims's 2003 book *Adam's Navel*.

15. Darwin 1872, 138.

16. Quoted in Isaac 1992, 58.

17. As I write, though, it looks as though the Australian cricketers may lose the Ashes series against England. *Note added later:* They did. *Note added even later* (2006/2007): Then they lost to them again. Now (late 2010) they look like losing again.

18. Calvin 1983.

19. Kirschmann 1999.

20. In that game, they actually don't seem to throw the ball, but sort of paddle it with the hand. But they kick the ball a lot, and kicking in this fashion, also done in rugby, itself requires a bipedal stance. As a New Zealander, I am tempted to think that pressure to kick a rugby ball was a factor in the evolution of bipedalism, but to my knowledge there is no evidence for rugby pitches dating from six million years ago, and in any event we had to wait for the Neandertals to evolve before we could pack a decent scrum.

21. Bowlers in cricket, for some obscure reason, are not allowed to flex the elbow, and compensate by running up to the bowling crease, thereby building up kinetic energy, before releasing the ball.

22. Bowling in cricket is again an exception, since fast bowlers build up speed before releasing the ball.

23. Marzke 1996.

24. Bronowski 1974, 115–116. Aristotle credited the Greek philosopher Anaxagoras with the view that it is because of the hands that humans are the wisest animal.

25. Young 2003.

26. Alemseged et al. 2006.

27. Westergaard et al. 2000.

28. Darwin 1896, 82.

29. Toth et al. 1993.

30. Bingham 1999.

31. Shaw 1948, 195.

32. The vote was taken in May 2009 by a committee of the International Commission on Stratigraphy (ICS)—see Kerr 2009.

33. Wood and Collard 1999.

34. Berger et al. 2010; Dirks et al. 2010.

35. Other members of the genus have been identified, but it depends a bit where you draw the lines of demarcation. That little group will do for now.

36. Foley 1984.

37. Bramble and Lieberman 2004.

38. Richard William Pearse, a New Zealand famer, is said to have taken to the air in a heavier-than-air machine on 31 March 1903, some nine months before the Wright brothers. The evidence on this, though, is still up in the air.

39. Darwin 1896, 64.

40. Hrdy 2009.

41. Another possibility would have been flight, but that was for the birds.

42. Tooby and DeVore 1987. An alternative view is that the third way did not arrive until Mr. Tony Blair was elected Prime Minister of Great Britain.

43. Humphrey 1976.

44. Alexander 1990, 4.

45. Wrangham 2009.

46. Goren-Inbar et al. 2004.

47. Or so it is said. Perhaps it applies especially to mammals, since crocodiles have very small brains, as did the dinosaurs.

48. They are bipedal, they fly, they learn complex vocal sequences, they make tools, and, as we saw in chapter 6, it has been claimed that they travel mentally in time.

49. Jerison 1973. The formula is EQ = (brainweight) / (.12 × bodyweight$^{.66}$). It is calibrated so that the EQ of the average mammal is 1.0, and the exponent of 0.66 compensates for the tendency of brain size to increase at a slower rate than body weight.

50. Every decimal point counts, it seems.

51. Dunbar 1993.

52. Her remains were discovered in 1976 by Donald Johanson and colleagues in Ethiopia, and named after the Beatles' song "Lucy in the Sky with Diamonds." Whether Lucy had ever used LSD is unknown. See Johanson and Edey 1981.

53. These figures are from Martin 1992.

54. Noonan et al. 2006.

55. Evans et al. 2004.

56. Mekel-Bobrov et al. 2005.

57. Evans et al. 2006.

58. For review, see Dorus et al. 2004.

59. In case you ask, the enzyme is called CMP-N-acetylneuraminic acid (CMP-Neu5Ac) hydroxylase (CMAH). The inactivating mutation of this gene has resulted in the absence in humans of the mammalian sialic acid N-glycolylneuraminic acid (Neu5Gc). This work is described by Chou et al. (2002).

60. Stedman et al. 2004.

61. Currie 2004.

62. It pains me to report that there is now some doubt as to the role of MYH16 in the increase in brain size—see McCollom et al. 2006.

63. Pinker 1994.

64. Actually, I don't really think it's a good idea to be rid of taxes, which have supported me throughout my academic life.

65. Deacon 1997.

66. Semendeferi, Damasio, and Frank (1997) showed that the ratio of the frontal lobes to the rest of the brain is constant in apes and humans, while Uylings (1990) suggests that the ratio has changed little from rats to humans. Deacon (1997, 476) is critical of these studies, partly on the grounds that they do not measure the prefrontal cortex independently of motor and premotor areas. A more recent study suggests that the volume of white matter in the prefrontal cortex is disproportionately large in humans relative to other primates, but the volume of grey matter is not (Schoenemann, Sheehan, and Glotzer 2005). Maybe white matter is what matters.

67. Flinn, Geary, and Ward 2005.

68. Dr. Darwin, that is.

69. Locke and Bogin 2006.

70. Locke and Bogin 2006, 262.

71. Busby Grant and Suddendorf 2009.

72. This is not to say that humans don't migrate seasonally. Wealthy Canadians migrate to Florida or Hawaii during the winter, and New Zealanders fly off to Queensland in Australia.

73. Anton 2002. One delightful exception is the hobbit-like hominin known as the Lady of Flora—or more formally as *Homo floresiensis*—whose skeleton remains were discovered on the island of Flora in Southeast Asia (Brown et al. 2004). She appears to date from a mere 18,000 years ago, which is by far the most recent date for any hominin who does not belong to our own species, *Homo sapiens*. An adult, she was only about one meter tall, with a brain volume of only about 380 cc, which is rather smaller than that of a modern chimpanzee. Her encephalization quotient (EQ) was nevertheless within the range 2.5–4.6, which compares with that of *Homo erectus* at between 3.3 and 4.4. The current consensus seems to be that she belonged indeed to the species *Homo erectus* (e.g., Falk et al. 2005), but had been subjected to what is known as "insular dwarfism" arising from long-term isolation and limited resources. As a New Zealander, I worry that the same thing might occur on our own islands and affect our rugby team—hobbits have already been detected in some areas.

74. Dennell and Roebroeks 2005.

75. Bogart and Pruetz 2008.

76. Pruetz and Bertolani 2007.

77. Boesch, Head, and Robbins 2009. The tools are identified as pounders, enlargers, collectors, perforators, and swabbers. They are mindful of the tools used in modern surgery.

78. Carvalho et al. 2009.

79. Sousa, Biro, and Matsuzawa 2009.

80. Beck 1980, 218.

81. Semaw et al. 1997.

82. Chazan et al. 2008.

83. Hunt 2000. These birds also shape twigs to act as hooks (Weir, Chappell, and Kacelnik 2002).

84. Walter et al. 2000.

85. Hoffecker 2007.

86. Noonan et al. 2006.

87. No one seems to have suggested that the horses made the spears.

88. Thieme 1997.

CHAPTER 12

Becoming Modern

1. As remarked in note 72 in chapter 4, it has become fairly apparent that *Homo sapiens* mated with the Neandertals some time after the migration from Africa, but before the Asian and European populations split—some 50,000 to

80,000 years ago. The estimated gene flow from Neandertal to non-African *H. sapiens* is estimated at between 1 and 4 percent (Green et al. 2010).

2. Atkinson, Gray, and Drummond 2009. Four other haplotypes have also been identified but these are rare.

3. Cann, Stoneking, and Wilson 1987. Tracing mtDNA back to a single woman does not mean that Mitochondrial Eve was the only woman alive at the time.

4. Atkinson, Gray, and Drummond 2008.

5. Mellars 2006a.

6. Brown et al. 2009. These authors suggest that the technological use of fire at Pinnacle Point in South Africa may go back as far as 164,000 years ago.

7. Goren-Inbar et al. 2004.

8. Henshilwood et al. 2002.

9. Mellars 2006b.

10. Carto et al. 2009.

11. Bowler et al. 2003.

12. Marean et al. 2007.

13. Mellars 2006a.

14. Crow 2010—see note 10 in chapter 4.

15. Hoffecker 2005, 195; emphasis added.

16. Crystal 1997.

17. Knight et al. 2003.

18. I suggested this in my 2002 book *From Hand to Mouth*, but that was before I knew about the *FOXP2* gene.

19. Especially the Italians.

20. Corballis 2004b.

21. Conard, Malina, and Münzel 2009.

22. Kvavadze et al. 2009.

23. Hoffecker 2005.

24. Conard 2009.

25. Mellars 2009, 177. He doesn't divulge the student's sex.

26. Mellars 2005.

27. Darwin 1896, 64.

28. Mellars and Stringer 1989.

29. Klein 2008. Again, of course, a candidate mutation is that of the *FOXP2* gene. As we saw in chapter 4, one recent estimate places this mutation at closer to 50,000 years ago than 100,000 years ago (Coop et al. 2008). Nevertheless this remains contentious, especially following evidence that the mutation may have been present in the common ancestor of human and Neandertal (Krause et al. 2007).

30. Powell, Shennan, and Thomas 2009. The estimated date of 45,000 years ago places this very early in the estimated date of migration into Europe, but there still seems to be some uncertainty in the dating of these events.

31. Ambrose 1998.

32. New Zealand is sometimes known as the shaky isles, especially in Australia, and an eruption of Lake Taupo in about AD 280 caused the sky to glow red as far away as Russia and China. It is said to have erupted some 28 times since the original huge eruption around 26,500 years ago, and it could pop again at any time.

33. Arthur 2007, 277.

34. Letter to Robert Hooke, dated 15 February 1676.

35. Although cultural variation is now well documented in chimpanzees, Whiten et al. (2009) suggest that, unlike human culture, chimpanzee culture is not cumulative. The ratchet-like nature of human culture allows it to cumulate over generations. This is perhaps another example of recursion, in which past cultural development is embedded in present culture.

36. Diamond 1997, 14. For "cargo" you can mostly read "junk"—stuff you don't really need.

Chapter 13

Final Thoughts

1. Chapter 13 is entitled "Secondary Sexual Characters of Birds."

2. Darwin 1896, 126. It is interesting that this passage begins with reference to higher animals and ends with reference to lower ones.

3. Arnold 1992.

4. E.g., Ambrose 2001.

5. There may nevertheless be cultural variation. The Pirahã, discussed in chapters 6 and 7, appear to have a relatively reduced language and sense of time, but may compensate in other ways. There is little question, moreover, that all humans possess the capacity for unlimited expression, but cultures vary to the manner in which that capacity is exploited.

6. Arthur 2007.

7. Dunbar 2004.

8. Me, I am scarcely able to even screw in a lightbulb.

9. Wiles 1995; Taylor and Wiles 1995. There remains the remote possibility that Fermat had a much simpler proof all along.

10. Hamilton 2005.

11. Beyerstein 1999.

12. My mother did try. She took me skiing at the age of 16 (my age, not hers), but conditions were fairly primitive, and after struggling some distance up the mountain I accidently released one ski, which sped all the way down again. I suppose if I had persevered with the remaining ski I might have invented snowboarding.

References

Addis, D. R., Wong, A. T. and Schacter, D. L. 2007. Remembering the past and imagining the future: Common and distinct neural substrates during event construction. *Neuropsychologia*, **45**, 1363–1377.

Aitken, P. G. 1981. Cortical control of conditioned and spontaneous vocal behaviour in rhesus monkeys. *Brain and Language*, **13**, 171–184.

Alemseged, Z., Spoor, F., Kimbel, W. H., Bobe, R., Geraads, D., Reed, D., et al. 2006. A juvenile early hominin skeleton from Dikika, Ethiopia. *Nature*, **443**, 296–301.

Alexander, R. D. 1990. *How did humans evolve? Reflections on the uniquely unique species*. Ann Arbor: Museum of Zoology, University of Michigan.

Ambrose, S. H. 1998. Late Pleistocene human population bottlenecks, volcanic winter, and the differentiation of modern humans. *Journal of Human Evolution*, **34**, 623–651.

———. 2001. Paleolithic technology and human evolution. *Science,* **291**, 1748–1753.

Andrews, P., and Stringer, C. 1993. The primates' progress. In S. J. Gould (ed.), *The book of life* (pp. 219–251). London: Norton.

Anton, S. C. 2002. Evolutionary significance of cranial variation in Asian *Homo erectus*. *American Journal of Physical Anthropology*, **118**, 310–323.

Arbib, M. A. 2005. From monkey-like action recognition to human language: An evolutionary framework for neurolinguistics. *Behavioral and Brain Sciences*, **28**, 105–168.

———. 2009. Invention and community in the emergence of language: A perspective from new sign languages. In S. M. Platek and T. K. Shackelford (eds.), *Foundations in cognitive neuroscience: Introduction to the discipline* (pp. 117–152). Cambridge: Cambridge University Press.

———. 2010. Mirror system activity for action and language is embedded in the integration of dorsal and ventral pathways. *Brain and Language*, **11**, 12–24.

Arbib, M. A., Liebal, K., and Pika, S. 2008. Primate vocalization, gesture, and the evolution of human language. *Current Anthropology*, **49**, 1053–1076.

Arbib, M. A., and Rizzolatti, G. 1997. Neural expectations: A possible evolutionary path from manual skill to language. *Communication and Cognition*, **29**, 393–424.

Arcadi, A. C. 2000. Vocal responsiveness in male wild chimpanzees: Implications for the evolution of language. *Journal of Human Evolution*, **39**, 205–223.

Arcadi, A. C., Robert, D., and Boesch, C. 1998. Buttress drumming by wild chimpanzees: Temporal patterning, phase integration into loud calls, and preliminary evidence for individual differences. *Primates*, **39**, 505–518.

Armstrong, D. F. 1999. *Original signs: Gesture, sign, and the source of language.* Washington, DC: Gallaudet University Press.

Armstrong, D. F., Stokoe, W. C., and Wilcox, S. E. 1995. *Gesture and the nature of language.* Cambridge: Cambridge University Press.

Armstrong, D. F., and Wilcox, S. E. 2007. *The gestural origin of language.* Oxford: Oxford University Press.

Arnold, V. I. 1992. *Catastrophe theory.* Berlin: Springer-Verlag.

Aronoff, M. 2007. In the beginning was the word. *Language*, **83**, 803–830.

Aronoff, M., Meir, I., Padden, C. A., and Sandler, W. 2008. The roots of linguistic organization in a new language. *Interaction Studies*, **9**, 133–153.

Arthur, W. B. 2007. The structure of invention. *Research Policy*, **36**, 274–287.

Aslin, R. N., Werker, J. F., and Morgan, J. L. 2002. Innate phonetic boundaries revisited (L). *Journal of the Acoustical Society of America*, **112**, 1258–1260.

Atance, C. M., and O'Neill, D. K. 2005. The emergence of episodic future thinking in humans. *Learning and Motivation*, **36**, 126–144.

Atkinson, Q. D., Gray, R. D., and Drummond, A. J. 2008. MtDNA variation predicts population size in humans and reveals a major southern Asian chapter in human prehistory. *Molecular Biological Evolution*, **25**, 468–474.

———. 2009. Bayesian coalescent inference of major human mitochondrial DNA haplogroup expansions in Africa. *Proceedings of the Royal Society B: Biological Sciences*, **276**, 367–373.

Austen, J. 1813. *Pride and prejudice.* London: T. Egerton.

Aziz-Zadeh, L., Wilson, S. M., Rizzolatti, G., and Iacoboni, M. 2006. Congruent embodied representations for visually presented actions and linguistic phrases describing actions. *Current Biology*, **16**, 1818–1823.

Badcock, C., and Crespi, B. 2006. Imbalanced genomic imprinting in brain development: An evolutionary basis for the aetiology of autism. *Journal of Evolutionary Biology*, **19**, 1007–1032.

Bahlmann, J., Gunter, T. C., and Friederici, A. D. 2006. Hierarchical and linear sequence processing: An electrophysiological exploration of two different grammar types. *Journal of Cognitive Neuroscience*, **18**, 1829–1842.

Barkow, J. H., Cosmides, L., and Tooby, J. (eds.) 1992. *The adapted mind: Evolutionary psychology and the generation of culture.* Oxford: Oxford University Press.

Baron-Cohen, S. 1995. *Mindblindness: An essay on autism and theory of mind.* Cambridge: A Bradford Book / MIT Press.

———. 2002. The extreme male brain theory of autism. *Trends in Cognitive Sciences*, **6**, 248–254.

———. 2009. Autism: The empathizing-systematizing (E-S) theory. In M. B. Miller and A. Kingstone (eds.), *The year in cognitive science 2009* (pp. 68–80). Boston: Blackwell.

Barth, J. 1969. *Lost in the funhouse.* London: Secker and Warburg.

Bartlett, F. C. 1932. *Remembering: A study in experimental and social psychology.* Cambridge: Cambridge University Press.

Bass, E., and Davis L. 1988. *The courage to heal: A guide for women survivors of child sexual abuse*. New York: Perennial Library.

Beck, B. B. 1980. *Animal tool behavior: The use and manufacture of tools by animals*. New York: Garland STPM Press.

Berger, L. R., de Ruiter, D. J., Churchill, S. E., Schmid, P. Carlson, K. J., Dirks, P.H.G.M., et al. 2010. *Australopithecus sediba*: A new species of *Homo*-like australopith from South Africa. *Science*, **328**, 195–204.

Bernard, L. L. 1924. *Instinct: A study in social psychology*. New York: Henry Holt.

Bernardis, P., Bello, A., Pettenati, P., Stefanini, S., and Gentilucci, M. 2008. Manual actions affect vocalizations of infants. *Experimental Brain Research*, **184**, 599–603.

Bernheim, H. 1989. *Suggestive therapeutics: A treatise on the nature and uses of hypnotism*. New York: Putnam.

Beyerstein, B. L. 1999. Whence cometh the myth that we only use 10% of our brains. In S. Della Sala (ed.), *Mind myths* (pp. 3–24). Chichester, UK: Wiley.

Bickerton, D. 1995. *Language and human behavior*. Seattle: University of Washington Press.

———. 2010. On two incompatible theories of language evolution. In R. K. Larson, V. Déprez, and H. Yamakido (eds.), *The evolution of human language* (pp. 199–210). Cambridge: Cambridge University Press.

Bingham, P. M. 1999. Human uniqueness: A general theory. *Quarterly Review of Biology*, **74**, 133–169.

Binkofski, F., and Buccino, G. 2004. Motor functions of the Broca's region. *Brain and Language*, **89**, 362–389.

Bischof, N. 1978. On the phylogeny of human morality. In G. S. Stent (ed.), *Morality as biological phenomenon* (pp. 53–74). Berlin: Dahlem Konferenzen.

Bischof-Köhler, D. 1985. Zur Phylogenese menschlicher Motivation [On the phylogeny of human motivation]. In L. H. Eckensberger and E. D. Lantermann (eds.), *Emotion und Reflexivität* (pp. 3–47). Vienna: Urban and Schwarzenberg.

Blatt, F. 1957. Latin influences on European syntax. *Travaux du Cercle Linguistique de Copenhague*, **11**, 33–69.

Bloom, P. 2004. *Descartes' baby: How the science of child development explains what makes us human*. New York: Basic Books.

Boë, L.-J., Heim, J.-L., Honda, K., Maeda, S., Badin, P., and Abry, C. 2007. The vocal tract of newborn humans and Neanderthals: Acoustic capabilities and consequences for the debate on the origin of language. A reply to Lieberman (2007a). *Journal of Phonetics*, **35**, 564–581.

Boë, L.-J., Honda, J.-L., Honda, K., and Maeda, S. 2002. The potential Neandertal vowel space was as large as that of modern humans. *Journal of Phonetics*, **30**, 465–484.

Boesch, C., and Boesch, H. 1990. Tool use and tool making in wild chimpanzees. *Folia Primatologica*, **54**, 86–99.

Boesch, C., Head, J., and Robbins, M. M. 2009. Complex tool sets for honey extraction among chimpanzees in Laongo National Park, Gabon. *Journal of Human Evolution*, 56, 560–569.

Bogart, S. L., and Pruetz, J. D. 2008. Ecological context of savanna chimpanzee (*Pan troglodyte verus*) termite fishing at Fongoli, Senegal. *American Journal of Primatology*, 70, 605–612.

Borke, H. 1975. Piaget's mountains revisited: Changes in the egocentric landscape. *Developmental Psychology*, 11, 240–243.

Botzung, A., Denkova, E., and Manning, L. 2008. Experiencing past and future personal events: Functional neuroimaging evidence on the neural basis of mental time travel. *Brain and Cognition*, 66, 202–212.

Bowler, J. M., Johnston, H., Olley, J. M., Prescott, J. R., Roberts, R. G., Shawcross, W., et al. 2003. New ages for human occupation and climate change at Lake Mungo, Australia. *Nature*, 421, 837–840.

Boyd, B. 2009. *The origin of stories: Evolution, cognition, and fiction*. Cambridge: Belnap Press of Harvard University Press.

Bramble, D. M., and Lieberman, D. E. 2004. Endurance running and the evolution of *Homo*. *Science*, 432, 345–352.

Bronowski, J. 1974. *The ascent of man*. Boston: Little, Brown.

Browman, C. P., and Goldstein, L. F. 1995. Dynamics and articulatory phonology. In T. van Gelder and R. F. Port (eds.), *Mind as motion* (pp. 175–193). Cambridge: MIT Press.

Brown, K. S., Marean, C. W., Herries, A. L. R., Jacobs, Z., Tribolo, C., Braun, D., et al. 2009. Fire as an engineering tool of early modern humans. *Science*, 325, 859–862.

Brown, P., Sutikna, T., Morwood, M. J., Soejono, R. P., Jatmiko, W. S., et al. 2004. A new small-bodied hominin from the late Pleistocene of Flores, Indonesia. *Nature*, 431, 1055–1061.

Bruce, D., Dolan, A., and Phillips-Grant, K. 2000. On the transition from childhood amnesia to the recall of personal memories. *Psychological Science*, 11, 360–364.

Brunet, M., Guy, F., Pilbeam, D., Mackaye, H. T., Likius, A., Ahounta, D., Beauvilain, A., et al. 2002. A new hominid from the Upper Miocene of Chad, Central Africa. *Nature*, 418, 145–151.

Burianova, H. and Grady, C. L. 2007. Common and unique activations in autobiographical, episodic, and semantic retrieval. *Journal of Cognitive Neuroscience*, 19, 1520–1534.

Burkart, J. M., and Heschl, A. 2007. Perspective taking or behaviour reading? Understanding of visual access in common marmosets (*Calithrix jacchus*). *Animal Behaviour*, 73, 457–469.

Burling, R. 1999. Motivation, conventionalization, and arbitrariness in the origin of language. In B. J. King (ed.), *The origins of language: What human primates can tell us* (pp. 307–350). Santa Fe, NM: School of American Research Press.

———. 2005. *The talking ape*. New York: Oxford University Press.

Burnham, W. H. 1989. Memory, historically and experimentally considered: III. Paramnesia. *American Journal of Psychology*, **2**, 568–622.

Busby Grant, J., and Suddendorf, T. 2009. Preschoolers begin to differentiate the times of events from throughout the lifespan. *European Journal of Developmental Psychology*, **6**, 746–762.

Butler, S. 1919. On the making of music, pictures and books. In H. F. Jones (ed.), *The note-books of Samuel Butler* (pp. 93–99). London: A.C. Fifield.

Butterfield, H. 1981. *The origins of history*. New York: Basic Books.

Byrne, R. W., and Whiten, A. 1990. Tactical deception in primates: The 1990 data base. *Primate Report*, **27**, 1–10.

Calvert, G. A., and Campbell, R. 2003. Reading speech from still and moving faces: The neural substrates of visible speech. *Journal of Cognitive Neuroscience*, **15**, 57–70.

Calvin, W. H. 1983. *The throwing Madonna: Essays on the brain*. New York: McGraw-Hill.

Calvin, W. H., and Bickerton, D. 2000. *Lingua ex machina: Reconciling Darwin and Chomsky with the human brain*. Cambridge: MIT Press.

Cann, R. L., Stoneking, M., and Wilson, A. C. 1987. Mitochondrial DNA and human evolution. *Nature*, **325**, 31–36.

Cargile, J. 1970. A note on "iterated knowings." *Analysis*, **30**, 151–155.

Carreiras, M., Lopez, J., Rivero, F., and Corina, D. 2005. Neural processing of a whistled language. *Nature*, **433**, 31–32.

Carto, S. L., Weaver, A. J., Heatherington, R., Lam, Y., and Wiebe, E. C. 2009. Out of Africa and into an ice age: On the role of global climate change in the late Pleistocene migration of early modern humans out of Africa. *Journal of Human Evolution*, **56**, 139–151.

Carvalho, S., Biro, D., McGrew, W. C., and Matsuzawa, T. 2009. Tool-composite reuse in wild chimpanzees (*Pan troglodytes*): Archaeologically invisible steps in the technological evolution of early hominins? *Animal Cognition*, **12** (Suppl. 1), 103.

Chase, S. 1938. *Tyranny of words*. London: Methuen.

Chazan, M., Ron, H., Matmon, A., Porat, N., Goldberg, P., Yates, R., et al. 2008. Radiometric dating of the earlier Stone Age sequence in Excavation I at Wonderwerk Cave, South Africa: Preliminary results. *Journal of Human Evolution*, **55**, 1–11.

Chen, J.-Y. 2007. Do Chinese and English speakers think about time differently? Failure of replicating Boroditsky 2001. *Cognition*, **104**, 427–436.

Cheney, D. L., and Seyfarth, R. S. 1990. *How monkeys see the world*. Chicago: University of Chicago Press.

Chomsky, N. 1957. *Syntactic structures*. The Hague: Mouton.

———. 1959. A review of Skinner's "Verbal Behavior." *Language*, **35**, 26–58.

———. 1966. *Cartesian linguistics: A chapter in the history of rationalist thought*. New York: Harper and Row.

———. 1970. *New horizons in the study of language and mind.* Cambridge: Cambridge University Press.

———. 1975. *Reflections on language.* New York: Pantheon.

———. 1988 *Language and the problem of knowledge: The Managua lectures.* Cambridge: MIT Press.

———. 1995. *The minimalist program.* Cambridge: MIT Press.

———. 2010. Some simple evo devo theses: How true might they be for language? In R. K. Larson, V. Déprez, and H. Yamakido, *The evolution of human language* (pp. 45–62). Cambridge: Cambridge University Press.

Chou, H.-H., Hakayama, T., Diaz, S., Krings, M., Indriati, E., Leakey, M., et al. 2002. Inactivation of CMP-N-acetylneuraminic acid hydroxylase occurred prior to brain expansion during human evolution. *Proceedings of the National Academy of Sciences* (USA), **99**, 11736–11741.

Christiansen, M. H., and Chater, N. 2008. Language as shaped by the brain. *Behavioral and Brain Sciences*, **31**, 489–558.

Christiansen, M. H., and Kirby, S. 2003. Language evolution: The hardest problem in science? In M. H. Christiansen and S. Kirby (eds.), *Language evolution* (pp. 1–15). Oxford: Oxford University Press.

Clayton, N. S., Bussey, T. J., and Dickinson, A. 2003. Can animals recall the past and plan for the future? *Trends in Cognitive Sciences*, **4**, 685–691.

Conard, N. J. 2009. A female figurine from the basal Aurignacian of Hohle Fels Cave in southwestern Germany. *Nature*, **459**, 248–252.

Conard, N. J., Malina, M., and Münzel, S. C. 2009. New flutes document the earliest musical tradition in southwestern Germany. *Nature*, **460**, 737–740.

Condillac, E. Bonnot de 1971. *An essay on the origin of human knowledge: Being a supplement to Mr. Locke's Essay on the human understanding.* A facsimile reproduction of the 1756 translation by T. Nugent of Condillac's 1747 essay. Gainesville, FL: Scholars' Facsimiles and Reprints.

Coop, G., Bullaughev, K., Luca, F., and Przeworski, M. 2008. The timing of selection of the human FOXP2 gene. *Molecular Biology and Evolution*, **25**, 1257–1259.

Corballis, M. C. 1991. *The lopsided ape.* New York: Oxford University Press.

———. 2002. *From hand to mouth: The origins of language.* Princeton, NJ: Princeton University Press.

———. 2004a. FOXP2 and the mirror system. *Trends in Cognitive Sciences*, **8**(2), 95–96.

———. 2004b. The origins of modernity: Was autonomous speech the critical factor? *Psychological Review*, **111**, 543–552.

———. 2007a. On phrase-structure and brain responses: A comment on Bahlmann, Gunter, and Friederici 2006. *Journal of Cognitive Neuroscience*, **19**, 1581–1583.

———. 2007b. Recursion, language, and starlings. *Cognitive Science*, **31**, 697–704.

Corballis, M. C., and Suddendorf, T. 2007. Memory, time, and language. In C. Pasternak (ed.), *What makes us human* (pp. 17–36). Oxford: Oneworld Publications.

Corkin, S. 2002. What's new with the amnesic patient H.M.? *Nature Reviews Neuroscience*, **3**, 153–160.

Correia, S. P. C., Dickinson, A., and Clayton, N. S. 2007. Western scrub-jays anticipate future needs independently of their current motivational state. *Current Biology*, **17**, 856–861.

Cosmides, L. 1985. The logic of social exchange: Has natural selection shaped how humans reason? Studies with the Wason selection task. *Cognition*, **31**, 187–276.

Cosmides, L., and Tooby, J. 1992. Cognitive adaptations for social exchange. In J. Barkow, L. Cosmides, and J. Tooby (eds.), *The adapted mind: Evolutionary psychology and the generation of culture* (pp. 163–228). New York: Oxford University Press.

Cox, N. 2004. *Tarzan Presley*. Wellington, New Zealand: Victoria University Press.

Crespi, B., and Badcock, C. 2008. Psychosis and autism as diametrical disorders of the human brain. *Behavioral and Brain Sciences*, **31**, 241–320.

Critchley, M. 1939. *The language of gesture*. London: Arnold.

———. 1975. *Silent language*. London: Arnold.

Crow, T. J. 2010. A theory of the origin of cerebral asymmetry: Epigenetic variation superimposed on a fixed right-shift. *Laterality*, **15**, 289–303.

Crystal, D. 1997. *The Cambridge encyclopedia of language*. 2nd edition. Cambridge: Cambridge University Press.

Culotta, E. 2007. Ancient DNA reveals Neandertals with red hair, fair complexions. *Science*, **318**, 546–547.

Currie, P. 2004. Muscling in on hominid evolution. *Nature*, **428**, 373–374.

Dally, J. M., Emery, N. J., and Clayton, N. S. 2006. Food-caching western scrub-jays keep track of who was watching when. *Science*, **312**, 1662–1666.

D'Argembeau, A., Xue, G., Lu, Z. L., Van der Linden, M., and Bechara, A. 2008. Neural correlates of envisioning emotional events in the near and far future. *NeuroImage*, **40**, 398–407.

Darwin, C. 1859. *On the origin of species by means of natural selection*. London: John Murray.

———. 1872. *The expression of the emotions in man and animals*. London: John Murray.

———. 1896. *The descent of man and selection in relation to sex*. 2nd edition. 1st edition published 1871. New York: Appleton.

Davis, P., and Kenyon, D. 2004. *Of pandas and people*. Richardson, TX: Foundation for Thought and Ethics.

Dawkins, R., and Krebs, J. R. 1978. Animal signals: Information or manipulation? In J. R. Krebs and N. B. Davies (eds.), *Behavioral ecology: An evolutionary approach* (pp. 283–309). Sunderland, MA: Sinauer.

Deacon, T. 1997. *The symbolic species*. New York: Norton.

de Boer, B., and Fitch, W. T. 2010. Computer models of vocal tract evolution: An overview and critique. *Adaptive Behavior, 18*, 36–47.

DeGusta, D., Gilbert, W. H., and Turner, S. P. 1999. Hypoglossal canal size and hominid speech. *Proceedings of the National Academy of Sciences, 96*, 1800–1804.

Dennell, R., and Roebroeks, W. 2005. An Asian perspective on early human dispersal from Africa. *Nature, 438*, 1099–1104.

Dennett, D. C. 1983. Intentional systems in cognitive ethology: The "Panglossian paradigm" defended. *Behavioral and Brain Science, 6*, 343–390.

———. 1995. *Darwin's dangerous idea: Evolution and the meaning of life*. New York: Simon and Schuster.

de Villiers, J. 2007. The interface of language and theory of mind. *Lingua, 117*, 1858–1878.

de Waal, F.B.M. 1982. *Chimpanzee politics: Power and sex among apes*. Baltimore: Johns Hopkins Press.

———. 2008. Putting the altruism back into altruism: The evolution of empathy. *Annual Review of Psychology, 59*, 279–300.

Diamond, J. 1997. *Guns, germs, and steel: The fates of human societies*. New York: Norton.

Dick, F., Bates, E., Wulfeck, B., Utman, J. A., Dronkers, N. F., and Gernsbacher, M. A. 2001. Language deficits, localization, and grammar: Evidence for a distributed model of language breakdown in aphasic patients and neurologically intact individuals. *Psychological Review, 108*, 759–788.

Dirks, P. H. G., Kibii, J. M., Kuhn, B. F., Steininger, C., Churchill, S. E., Kramers, J. D., et al. 2010. Geological setting and age of Australopithecus sediba from southern Africa. *Science, 328*, 205–208.

Dobzhansky, T. 1973. Nothing in biology makes sense except in the light of evolution. *American Biology Teacher, 35*, 125–129.

Donald, M. 1991. *Origins of the modern mind*. Cambridge: Harvard University Press.

Dorus, S., Vallender, E. J., Evans, P. D., Anderson, J. R., Gilbert, S. L., Mahowald, M., et al. 2004. Accelerated evolution of nervous system genes in the origin of *Homo sapiens*. *Cell, 119*, 1027–1040.

Dostoevsky, F. 2008. *Notes from underground*. New York: Classic House (originally published 1864).

Dunbar, R.I.M. 1993. Coevolution of neocortical size, group size, and language in humans. *Behavioral and Brain Sciences, 16*, 681–735.

———. 2004. *The human story: A new history of mankind's evolution*. London: Faber and Faber.

Eckstein, G. 1948. *Everyday miracle*. New York: Harper and Row.

———. 1970. *The body has a head*. New York: Harper and Row.

Emmorey, K. 2002. *Language, cognition, and brain: Insights from sign language research*. Hillsdale, NJ: Erlbaum.

Enard, W., Gehre, S., Hammerschmidt, K., Halter, S. M., Blass, T., Somel, M., et al. 2009. A humanized version of Foxp2 affects cortico-basal ganglia circuits in mice. *Cell,* 137, 961–971.

Enard, W., Przeworski, M., Fisher, S. E., Lai, C.S.L., Wiebe, V., Kitano, T., et al. 2002. Molecular evolution of FOXP2, a gene involved in speech and language. *Nature,* 418, 869–871.

Esslinger, C., Walter, H., Kirsch, P., Erk, S., Schnell, K., Arnold, C., et al. 2009. Neural mechanisms of a genome-wide supported psychosis variant. *Science,* 324, 605.

Evans, N. 2003. *Bininj Gun-wok: A pan-dialectical grammar of Mayali, Kunwinjku and Kune*. Canberra, Australia: Pacific Linguistics.

———. 2009. *Dying words: Endangered languages and what they have to tell us*. Oxford: Wiley-Blackwell.

Evans, N., and Levinson, S. C. 2009. The myth of language universals: Language diversity and its importance for cognitive science. *Behavioral & Brain Sciences,* 32, 429–492.

Evans, P. D., Anderson, J. R., Vallender, E. J., Gilbert, S. L., Malcom, C. M., Dorus, S., et al. 2004. Adaptive evolution of ASPM, a major determinant of cerebral cortical size in humans. *Human Molecular Genetics,* 13, 489–494.

Evans, P. D., Mekel-Bobrov, N., Yallender, E. J., Hudson, R. R., and Lahn, B. T. 2006. Evidence that the adaptive allele of the brain size gene *microcephalin* introgressed into *Homo sapiens* from an archaic *Homo* lineage. *Proceedings of the National Academy of Sciences,* 103, 18178–18183.

Everett, D. L. 2005. Cultural constraints on grammar and cognition in Pirahã. *Current Anthropology,* 46, 621–646.

———. 2007. *Cultural constraints on grammar in Pirahã: A reply to Nevins, Pesetsky, and Rodrigues (2007)*. Online at http://ling.auf.net/lingBuzz/000427.

———. 2008. *Don't sleep, there are snakes*. New York: Pantheon.

Fadiga, L., Fogassi, L., Pavesi, G., and Rizzolatti, G. 1995. Motor facilitation during action observation: A magnetic stimulation study. *Journal of Neurophysiology,* 73, 2608–2611.

Falk, D., Hildebolt, C., Smith, K., Morwood, M. J., Sutkina, T., Brown, P., et al. 2005. The brain of LB1, *Homo floresiensis*. *Science,* 308, 242–245.

Farmelo, G. 2009. *The strangest man: The hidden life of Paul Dirac*. London: Faber and Faber.

Fauconnier, G. 2003. Cognitive linguistics. In L. Nadel (ed.), *Encyclopedia of cognitive science*, vol. 1 (pp. 539–543). London: Nature Publishing Group.

Ferkin, M. H., Combs, A., delBarco-Trillo, J., Pierce, A. A., and Franklin, S. 2008. Meadow voles, *Microtus pennsylvanicus*, have the capacity to recall the "what", "where", and "when" of a single past event. *Animal Cognition,* 11, 147–159.

Fisher, S. E., and Scharff, C. 2009. FOXP2 as a molecular window into speech and language. *Trends in Genetics*, 25, 166–177.

Fisher, S. E., Vargha-Khadem, F., Watkins, K. E., Monaco, A. P., and Pembrey, M. E. 1998. Localisation of a gene implicated in a severe speech and language disorder. *Nature Genetics,* 18, 168–170.

Fitch, W. T. 2010a. *The evolution of language.* Cambridge: Cambridge University Press.

———. 2010b. Three meanings of "recursion": Key distinctions for biolinguistics. In R. K. Larson, V. Déprez, and H. Yamakido (eds.), *The evolution of human language* (pp. 73–90). Cambridge: Cambridge University Press.

Fitch, W. T., and Hauser, M. D. 2004. Computational constraints on syntactic processing in a nonhuman primate. *Science, 303,* 377–380.

Flinn, M. V., Geary, D. C., and Ward, C. V. 2005. Ecological dominance, social competition, and the coalitionary arms race: Why humans evolved extraordinary intelligence. *Evolution and Human Behavior,* 26, 10–46.

Fodor, J. A. 1975. *The language of thought.* New York: Crowell.

———. 1983. *The modularity of mind.* Cambridge: Bradford Books, MIT Press.

———. 2000. *The mind doesn't work that way.* Cambridge: Bradford Books, MIT Press.

Foley, R. 1984. Early man and the Red Queen: Tropical African community evolution and hominid adaptation. In R. Foley (ed.), *Hominid evolution and community ecology: Prehistoric human adaptation in biological perspective* (pp. 85–110). London: Academic Press.

Forster, M. 1995. *Hidden lives.* New York: Viking.

Frege, G. 1980. Letter to Jourdain. In G. Gabriel, H. Hermes, F. Kambartel, C. Thiel, and A. Veraart (eds.), *Philosophical and mathematical correspondence* (pp. 78–80). Chicago: University of Chicago Press (originally published 1914).

Friederici, A. D., Bahlmann, J., Heim, S., Schubotz, R. I., and Anwander, A. 2006. The brain differentiates human and non-human grammars: Functional localization and structural connectivity. *Proceedings of the National Academy of Sciences,* 103, 2458–2463.

Frishberg, N. 1975. Arbitrariness and iconicity in American Sign Language. *Language,* 51, 696–719.

Galik, K., Senut, B., Pickford, M., Gommery, D., Treil, J., Kuperavage, A. J., et al. 2004. External and internal morphology of the BAR 1002'00 *Orrorin tugenensis* femur. *Science,* 305, 1450–1453.

Gallup, G. G., Jr. 1998. Self-awareness and the evolution of social intelligence. *Behavioural Processes,* 42, 239–247.

Gardner, R. A., and Gardner, B. T. 1969. Teaching sign language to a chimpanzee. *Science,* 165, 664–672.

Gentilucci, M., Benuzzi, F., Gangitano, M., and Grimaldi, S. 2001. Grasp with hand and mouth: A kinematic study on healthy subjects. *Journal of Neurophysiology,* 86, 1685–1699.

Gentilucci, M., and Corballis, M. C. 2006. From manual gesture to speech: A gradual transition. *Neuroscience and Biobehavioral Reviews*, 30, 949–960.

Gentner, T. Q., Fenn, K. M., Margoliash, D., and Nusbaum, H. C. 2006. Recursive syntactic pattern learning by songbirds. *Nature,* 440, 1204–1207.

Gibbons, A. 2009. A new kind of ancestor: *Ardipithecus* unveiled. *Science*, 326, 36–40.

Gibbs, R. W. 2000. Irony in talk among friends. *Metaphor and Symbol*, 15, 5–27.

Givòn, T. 1971. Historical syntax and synchronic morphology: An archaeologist's field trip. *CLS#7*, University of Chicago, Chicago Linguistics Society.

Goldin-Meadow, S., and McNeill, D. 1999. The role of gesture and mimetic representation in making language the province of speech. In M. C. Corballis and S. E. G. Lea (eds.), *The descent of mind* (pp. 155–172). Oxford: Oxford University Press.

Goodall, J. 1986. *The chimpanzees of Gombe: Patterns of behavior*. Cambridge: Harvard University Press.

Goren-Inbar, N., Alperson, N., Kislev, M. E., Simchoni, O., Melamed, Y., Ben-Nun, A., et al. 2004. Evidence for hominin control of fire at Gesher Benot, Ya'aqov, Israel. *Science*, 304, 725–727.

Grandin, T. 1996. *Thinking in pictures*. New York: Vintage.

Grandin, T., and Barron, S. 2005. *The unwritten rules of social relationships*. Arlington, TX: Future Horizons.

Grandin, T., and Johnson, C. 2005. *Animals in translation: Using the mysteries of autism to decode animal behavior*. New York: Scribner.

Grandin, T., and Scariano, M. M. 1986. *Emergence, labeled autistic*. Norato, CA: Arena Press.

Green, R. E., Krause, J., Briggs, A. W., Maricic, T., Stenzel, U., Kircher, M., et al. 2010. A draft sequence of the Neandertal genome. *Science*, 328, 710–722.

Greenfield, S., and Savage-Rumbaugh, E. S. 1990. Grammatical combination in *Pan paniscus*: Processes of learning and invention in the evolution and development of language. In S. T. Parker and K. R. Gibson (eds.), *"Language" and intelligence in monkeys and apes* (pp. 540–578). Cambridge: Cambridge University Press.

Grice, H. P. 1975. Logic and conversation. In P. Cole and J. Morgan (eds.), *Syntax and semantics*, vol. 3, *Speech acts* (pp. 43–58). New York: Academic Press.

———. 1989. *Studies in the way of words*. Cambridge: Harvard University Press.

Griffin, D. R. 2001. *Animal minds: Beyond cognition to consciousness*. Chicago: University of Chicago Press.

Grodzinsky, Y. 2006. The language faculty, Broca's region, and the mirror system. *Cortex*, 42, 464–468.

Gross, C. G. 1993. Huxley versus Owen: The hippocampus minor and evolution. *Trends in Neurosciences*, 16, 493–498.

Groszer, M., Keays, D. A., Deacon, R.M.J., de Bono, J. P., Prasad-Mulcare, S.,

Gaube, S., et al. 2008. Impaired synaptic plasticity and motor learning in mice with a point mutation implicated in human speech deficits. *Current Biology*, **18**, 354–362.

Haesler, S., Rochefort, C., Georgi, B., Licznerski, P., Osten, P., and Scharff, C. 2007. Incomplete and inaccurate vocal imitation after knockdown of FOXP2 in songbird basal ganglia nucleus area X. *PLoS Biology*, **5**, 2885–2897.

Hamilton, W. D. 2005. *Narrow roads of the gene land*, vol. 3, *Last words*. New York: Oxford University Press.

Happé, F. 1995. Understanding minds and metaphors: Insights from the study of figurative language in autism. *Metaphor and Symbolic Activity*, **10**, 275–295.

Harcourt, A. H., and Stewart, K. J. 2007. Gorilla society: What we know and don't know. *Evolutionary Anthropology*, **16**, 147–158.

Hare, B., and Tomasello, M. 1999. Domestic dogs (*Canis familiaris*) use human and conspecific social cues to locate hidden food. *Journal of Comparative Psychology*, **113**, 173–177.

Hare, B., Call, J., Agnetta, B., and Tomasello, M. 2000. Chimpanzees know what conspecifics do and do not see. *Animal Behaviour*, **59**, 771–785.

Hare, B., Call, J., and Tomasello, M. 2001. Do chimpanzees know what conspecifics know? *Animal Behaviour*, **61**, 139–151.

———. 2006. Chimpanzees deceive a human competitor by hiding. *Cognition*, **101**, 495–154.

Hassabis, D., Kumaran, D., and Maguire, E. A. 2007. Using imagination to understand the neural basis of episodic memory. *Journal of Neuroscience*, **27**, 14365–14374.

Hauser, M., and Carey, S. 1998. Building a cognitive creature from a set of primitives. In D. D. Cummins and C. Allen (eds.), *The evolution of mind* (pp. 51–106). New York: Oxford University Press.

Hauser, M. D., Chomsky, N., and Fitch, W. T. 2002. The faculty of language: What is it, who has it, and how did it evolve? *Science*, **298**, 1569–1579.

Hayakawa, S. I. 1972. *Language in thought and action*. New York: Harcourt Brace Jovanovich.

Hayes, C. 1952. *The ape in our house*. London: Gollancz.

Heine, B., and Kuteva, T. 2007. *The genesis of grammar*. Oxford: Oxford University Press.

Hennessy, P. 1995. *The hidden wiring: Unearthing the British constitution*. London: Gollancz.

Henry, O. 1910. *The ransom of the red chief*. Online at www.literaturecollection.com/a/o_henry/3/.

Henshilwood, C. S., d'Errico, F., Yates, R., Jacobs, Z., Tribolo, C., Duller, G.A.T., et al. 2002. Emergence of modern human behaviour: Middle Stone Age engravings from South Africa. *Science*, **295**, 1278–1280.

Herman, L. M., Richards, D. G., and Wolz, J. P. 1984. Comprehension of sentences by bottle-nosed dolphins. *Cognition*, **16**, 129–219.

Hewes, G. W. 1973. Primate communication and the gestural origins of language. *Current Anthropology,* **14,** 5–24.

Hickok, G. S. 2009. Eight problems for the mirror neuron theory of action understanding in monkeys and humans. *Journal of Cognitive Neuroscience,* **21,** 1229–1243.

Hihara, S., Yamada, H., Iriki, A., and Okanoya, K. 2003. Spontaneous vocal differentiation of coo-calls for tools and food in Japanese monkeys. *Neuroscience Research,* **45,** 383–389.

Hockett, C. F. 1960. The origins of speech. *Scientific American,* **203**(3), 88–96.

———. 1978. In search of love's brow. *American Speech,* **53,** 243–315.

Hodges, J. R., and Graham, K. S. 2001. Episodic memory: Insights from semantic dementia. *Philosophical Transactions of the Royal Society of London B: Biological Sciences,* **356,** 1423–1434.

Hoffecker, J. F. 2005. Innovation and technological knowledge in the Upper Paleolithic of northern Eurasia. *Evolutionary Anthropology,* **14,** 186–198.

———. 2007. Representation and recursion in the archaeological record. *Journal of Archaeological Method and Theory,* **14,** 359–387.

Hood, B. M. 2009. *Why we believe in the unbelievable: From superstition to religion—the brain science of belief.* San Francisco: HarperOne.

Hood, L. 2001. *A city possessed: The Christchurch Civic Crèche case.* Dunedin, New Zealand: Longacre Press.

Hopkins, W. D., Taglialatela, J. P., and Leavens, D. 2007. Chimpanzees differentially produce novel vocalizations to capture the attention of a human. *Animal Behaviour,* **73,** 281–286.

Hopper, P. J., and Traugott, E. C. 2003. *Grammaticalization.* 2nd edition. Cambridge: Cambridge University Press.

Horne Tooke, J. 1857. *Epea pteroenta or the diversions of Purley.* London.

Horrobin, D. 2003. *The madness of Adam and Eve: Did schizophrenia shape humanity?* London: Bantam Press.

Hrdy, S. B. 2009. *The evolutionary origins of mutual understanding.* Cambridge: Harvard University Press.

Humphrey, N. K. 1976. The social function of intellect. In P. P. G. Bateson and R. A. Hinde (eds.), *Growing points in ethology* (pp. 303–318). New York: Cambridge University Press.

Hunt, G. R. 2000. Human-like, population-level specialization in the manufacture of pandanus tools by New Caledonian crows Corvus moneduloides. *Proceedings of the Royal Society of London, Series B,* **267,** 403–413.

Hunt, G. R., and Gray, R. D. 2003. Diversification and cumulative evolution in New Caledonian crow tool manufacture. *Proceedings of the Royal Society of London, Series B,* **270,** 867–874

Ingvar, D. H. 1979. Hyperfrontal distribution of the cerebral grey matter flow in resting wakefulness: On the functional anatomy of the conscious state. *Neurological Scandinavica,* **60,** 12–25.

Inoue, S., and Matsuzawa, T. 2007. Working memory of numerals in chimpanzees. *Current Biology*, **17**(23), R1004–R1005.

Isaac, B. 1992. Throwing. In S. Jones, R. Martin, and D. Pilbeam (eds.), *The Cambridge Encyclopedia of human evolution* (p. 58). Cambridge: Cambridge University Press.

Jackendoff, R. 2002. *Foundations of language: Brain, meaning, grammar, evolution*. Oxford: Oxford University Press.

James, W. 1910. *Psychology: The briefer course*. New York: Holt.

Jarvis, E. D. 2006. Selection for and against vocal learning in birds and mammals. *Ornithological Science*, **5**, 5–14.

Jerison, H. J. 1973. *Evolution of brain and intelligence*. New York: Academic Press.

Johanson, D., and Edey, M. 1981. *Lucy: The beginnings of humankind*. New York: Simon and Schuster.

Johnson, F. 2000. I'm a soul man. *The Spectator*, **284** (N. 89581), 10–11.

Jürgens, U. 2002. Neural pathways underlying vocal control. *Neuroscience and Biobehavioral Reviews,* **26**, 235–258.

Kamil, A. C., and Balda, R. P. 1985. Cache recovery and spatial memory in Clark's nutcrackers (*Nucifraga columbiana*). *Journal of Experimental Psychology: Animal Behavior Processes*, **85**, 95–111.

Kaminski, J., Call, J., and Fischer, J. 2004. Word learning in a domestic dog: Evidence for "fast mapping." *Science*, **304**, 1682–1683.

Karlsson, F. 2007. Constraints on multiple center-embedding of clauses. *Journal of Linguistics*, **43**, 365–392.

Kaufman, E. L., Lord, E. W., Reese, M. W., and Volkman, J. 1949. The discrimination of visual number. *American Journal of Psychology*, **62**, 498–525.

Kay, R. F., Cartmill, M., and Barlow, M. 1998. The hypoglossal canal and the origin of human vocal behavior. *Proceedings of the National Academy of Sciences*, **95**, 5417–5419.

Kenneally, C. 2007. *The first word: The search for the origins of language*. Oxford: Oxford University Press.

Kéri, S. 2009. Genes for psychosis and creativity. *Psychological Science*, **20**, 1070–1073.

Kerr, R. A. 2009. The quaternary period wins out in the end. *Science*, **324**, 1249.

Kingsley, M. 1897. *Travels in West Africa, Cong Française, Corisco, and Cameroons*. London: F. Cass (reprinted 1965).

Kirby, S., and Hurford, J. R. 2002. The emergence of linguistic structure: An overview of the iterated learning model. In A. Cangelosi and D. Parisi (eds.), *Simulating the evolution of language* (pp. 121–148). London: Springer Verlag.

Kirschmann, E. 1999. *Das Zeitalter der Werfer*. Hannover, Germany: Eduard Kirschmann Grunlinde 4, 30459.

Klein, R. G. 2008. Out of Africa and the evolution of human behaviour. *Evolutionary Anthropology*, **17**, 267–281.

Klein, S. B., Loftus, J., and Kihlstrom, J. F. 2002. Memory and temporal experience: The effects of episodic memory loss on an amnesiac patient's ability to remember the past and imagine the future. *Social Cognition*, **20**, 353–379.

Klima, S. E., and Bellugi, U. 1979. *The signs of language*. Cambridge: Harvard University Press.

Knight, A., Underhill, P. A., Mortensen, H. M., Zhivotovsky, L. A., Lin, A. A., Henn, B. M., et al. 2003. African Y chromosome and mtDNA divergence provides insight into the history of click languages. *Current Biology*, **13**, 464–473.

Knight, C. 1998. Ritual/speech coevolution: A solution to the problem of deception. In J. R. Hurford, M. Studdert-Kennedy, and C. Knight (eds.), *Approaches to the evolution of language* (pp. 68–91). Cambridge: Cambridge University Press.

Kohler, E., Keysers, C., Umilta, M. A., Fogassi, L., Gallese, V., and Rizzolatti, G. 2002. Hearing sounds, understanding actions: Action representation in mirror neurons. *Science*, **297**, 846–848.

Köhler, W. 1925. *The mentality of apes*. New York: Routledge and Kegan Paul. (Originally published in German in 1917).

Konner, M. 1982. *The tangled wing: Biological constraints on the human spirit*. New York: Harper.

Krause, J., Lalueza-Fox, C., Orlando, L., Enard, W., Green, R. E., Burbano, H. A., et al. 2007. The derived FOXP2 variant of modern humans was shared with Neandertals. *Current Biology*, **17**, 1908–1912.

Kundera, M. 2002. *Ignorance*. New York: HarperCollins (translated by L. Asher).

Kvavadze, E., Bar-Yosef, O., Belfer-Cohen, A., Boaretto, E., Jakeli, N., Matskevich, Z., et al. 2009. 30,000-year-old flax fibers. *Science*, **325**, 1359.

Ladygina-Kohts, N. N. 2002. *Infant chimpanzee and human child*. Oxford: Oxford University Press. (Translated from the 1935 Russian version by B. Vekker.)

Lai, C. S., Fisher, S. E., Hurst, J. A., Vargha-Khadem, F., and Monaco A. P. 2001. A novel forkhead-domain gene is mutated in a severe speech and language disorder. *Nature*, **413**, 519–523.

Langford, D. J., Crager, S. E., Shehzad, Z., Smith, S. B., Sotocinal, S. G., et al. 2006. Social modulation of pain as evidence for empathy in mice. *Science*, **312**, 1967–1970.

Leakey, M. D. 1979. Footprints in the ashes of time. *National Geographic*, **155**, 446–457.

Leslie, A. M. 1994. ToMM, ToBY, and agency: Core architecture and domain specificity. In L. A. Hirschfeld and S. A. Gelman (eds.), *Mapping the mind: Domain specificity in cognition and culture* (pp. 119–148). New York: Cambridge University Press.

Liberman A. M., Cooper F. S., Shankweiler, D. P., and Studdert-Kennedy, M. 1967. Perception of the speech code. *Psychological Review*, **74**, 431–461.

Liebal, K., Call, J., and Tomasello, M. 2004. Use of gesture sequences in chimpanzees. *American Journal of Primatology,* **64**, 377–396.

Lieberman, D. E. 1998. Sphenoid shortening and the evolution of modern cranial shape. *Nature,* **393**, 158–162.

Lieberman, D. E., McBratney, B. M., and Krovitz, G. 2002. The evolution and development of cranial form in *Homo sapiens. Proceedings of the National Academy of Sciences,* **99**, 1134–1139.

Lieberman, P. 1998. *Eve spoke: Human language and human evolution.* New York: Norton.

———. 2007. The evolution of human speech. *Current Anthropology,* **48**, 39–46.

Lieberman, P., Crelin, E. S., and Klatt, D. H. 1972. Phonetic ability and related anatomy of the new-born, adult human, Neanderthal man, and the chimpanzee. *American Anthropologist,* **74**, 287–307.

Liégeois, F., Baldeweg, T., Connelly, A., Gadian, D. G., Mishkin, M., and Vargha-Khadem, F. 2003. Language fMRI abnormalities associated with FOXP2 gene mutation. *Nature Neuroscience,* **6**, 1230–1237.

Lin, J.-W. 2006. Time in a language without tense: The case of Chinese. *Journal of Semantics,* **23**, 1–53.

Liszkowski, U., Schafer, M., Carpenter, M., and Tomasello, M. 2009. Prelinguistic infants, but not chimpanzees, communicate about absent entities. *Psychological Science,* **20**, 654–660.

Littlewood, J. E. 1960. *A mathematician's miscellany.* London: Methuen.

Locke, J. L., and Bogin, B. 2006. Language and life history: A new perspective on the development and evolution of human language. *Behavioral and Brain Sciences,* **29**, 259–325.

Loftus, E., and Ketcham, K. 1994. *The myth of repressed memory.* New York: St. Martin's Press.

Loftus, E. F., and Loftus, G. R. 1980. On the permanence of stored information in the human-brain. *American Psychologist,* **35**, 409–420.

Lotto, A. J., Hickok, G. S., and Holt, L. L. 2009. Reflections on mirror neurons and speech perception. *Trends in Cognitive Sciences,* **13**, 110–114.

Lovejoy, C. O., Suwa, G., Spurlock, L., Asfaw, B., and White, T. D. 2009. The pelvis and femur of *Ardipithecus ramidus*: The emergence of upright walking. *Science,* **326**, 71, 71e1–71e6.

Luria, A. R. 1968. *The mind of a mnemonist: A little book about a vast memory.* London: Basic Books.

Luyster, R. J., Kadlee, M. B., Carter, A., and Tager-Flusberg, H. 2008. Language assessment in toddlers with autism spectrum disorders. *Journal of Autism and Developmental Disorders,* **38**, 1426–1438.

Mackenzie, D. 2009. A ticket, a tasket, an Apollonian gasket. *American Scientist,* **98**(6), 10–14.

MacLarnon, A., and Hewitt, G. 2004. Increased breathing control: Another factor in the evolution of human language. *Evolutionary Anthropology,* **13**, 181–197.

Maclean, P. D., and Newman, J. D. 1988. Role of midline frontolimbic cortex in production of isolation calls of squirrel-monkeys. *Brain Research*, **450**, 111–123.

MacNeilage, P. N. 2008. *The origin of speech*. Oxford: Oxford University Press.

Malotki, E. 1983. *Hopi time: A linguistic analysis of the temporal concepts in the Hopi language*. Berlin: Mouton.

Marchant, J. 1916. *Alfred Russel Wallace: Letters and reminiscences*. London: Cassell.

Marean, C. W., Bar-Matthews, M. Bernatchez, J., Fisher, E., Goldberg, P., Herries, A.I.R., et al. 2007. Early human use of marine resources and pigment in South Africa during the Middle Pleistocene. *Nature*, **449**, 905–1011.

Marks, D. F. 2000. *The psychology of the psychic*. 2nd edition. New York: Prometheus Books.

Marks, D. F., and Kammann, R. 1980. *The psychology of the psychic*. Amherst, NY: Prometheus Books.

Markus, H., and Nurius, P. 1986. Possible selves. *American Psychologist*, **41**, 954–969.

Marler, P. 1991. The instinct to learn. In S. Carey and B. Gelman (eds.), *The epigenesis of mind: Essays on biology and cognition* (pp. 37–66). Hillsdale NJ: Erlbaum.

Marshall, J., Atkinson, J., Smulovitch, E., Thacker, A., and Woll, B. 2004. Aphasia in a user of British Sign Language: Dissociation between sign and gesture. *Cognitive Neuropsychology*, **21**, 537–554.

Martin, R. 1992. Classification and evolutionary relationships. In S. Jones, R. Martin, and D. Pilbeam (eds.), *The Cambridge encyclopedia of human evolution* (pp. 17–23). Cambridge: Cambridge University Press.

Marzke, M. W. 1996. Evolution of the hand and bipedality. In A. Lock and C. R. Peters (eds.), *Handbook of symbolic evolution* (pp. 126–154). Oxford: Oxford University Press.

Maudsley, H. 1873. *Body and mind: An inquiry into their connection and mutual influence*. 2nd edition. London: Macmillan.

McCollom, M. A., Sherwood, C. C., Vinyard, C. J., Lovejoy, C. O., and Schachat, F. 2006. Of muscle-bound crania and human brain evolution: The story behind the *MYH16* headlines. *Journal of Human Evolution*, **50**, 232–236.

McDougall, W. 1908. *An introduction to social psychology*. London: Methuen.

McGrew, W. C. 2010. Chimpanzee technology. *Science*, **328**, 579–580.

McGurk, H., and MacDonald, J. 1976. Hearing lips and seeing voices. *Nature*, **264**, 746–748.

McNeill, D. 1992. *Hand and mind: What gestures reveal about thought*. Chicago: University of Chicago Press.

Mekel-Bobrov, N., Gilbert, S. L., Evans, P. D., Vallender, E. J., Anderson, J. R., Hudson, R. R., et al. 2005. On-going adaptive evolution of *ASPM*, a brain size determinant in *Homo sapiens*. *Science*, **309**, 1720–1722.

Mellars, P. A. 2005. The impossible coincidence: A single-species model for the origins of modern human behavior in Europe. *Journal of Human Evolution*, **14**, 12–27.

———. 2006a. Going east: New genetic and archaeological perspectives on the modern colonization of Eurasia. *Science*, **313**, 796–800.

———. 2006b. Why did modern human populations disperse from Africa ca. 60,000 years ago? A new model. *Proceedings of the National Academy of Sciences*, **103**, 9381–9386.

———. 2009. Origins of the female image. *Nature*, **459**, 176–177.

Mellars, P. A., and Stringer, C. B. (eds.) 1989. *The human revolution: Behavioral and biological perspectives on the origins of modern humans*. Edinburgh: Edinburgh University Press.

Miles, H. L. 1990. The cognitive foundations for reference in a signing orangutan. In S. T. Parker and K. R. Gibson (eds.), *Language and intelligence in monkeys and apes* (pp. 511–539). Cambridge: Cambridge University Press.

Mitchell, D. B. 2006. Nonconscious priming after 17 years. *Psychological Science*, **17**, 925–929.

Mithen, S. 1996. *The prehistory of the mind*. London: Thames and Hudson.

———. 2005. *The singing Neanderthals: The origins of music, language, mind, and body*. London: Weidenfeld and Nicholson.

Morell, V. 2009. Going to the dogs. *Science*, **325**, 1062–1065.

Moscovitch, M., Nadel, L., Winocur, G., Gilboa, A., and Rosenbaum, R. S. 2006. The cognitive neuroscience of remote episodic, semantic and spatial memory. *Current Opinion in Cognitive Neuroscience*, **16**, 179–190.

Müller, F. M. 1873. Lectures on Mr Darwin's philosophy of language. *Frazer's Magazine* 7 and 8. Reprinted in R. Harris (ed.) 1996, *The origin of language* (pp. 147–233). Bristol: Thoemmes Press.

Muir, L. J., and Richardson, I. E. G. 2005. Perception of sign language and its application to visual communication for deaf people. *Journal of Deaf Studies and Deaf Education*, **10**, 390–401.

Mulcahy, N. J., and Call, J. 2006. Apes save tools for future use. *Science*, **312**, 1038–1040.

Myers, E. B., Blumstein, S. E., Walsh, E., and Eliassen, J. 2009. Inferior frontal regions underlie the perception of phonetic category invariance. *Psychological Science*, **20**, 895–903.

Neidle, C., Kegl, J., MacLaughlin, D., Bahan, B., and Lee, R. G. 2000. *The syntax of American Sign Language*. Cambridge: MIT Press.

Neisser, U. 2008. Memory with a grain of salt. In H. H. Wood and A. S. Byatt (eds.), *Memory: An anthology* (pp. 80–88). London: Chatto and Windus.

Nevins, A., Pesetsky, D., and Rodrigues, C. 2007. *Pirahã exceptionality: A reassessment*. Online at http://ling.auf.net/lingBuzz/000411.

New, J., Cosmides, L., and Tooby, J. 2007. Category-specific attention for animals reflects ancestral priorities, not expertise. *Proceedings of the National Academy of Sciences*, **104**, 16598–16603.

Newton, M. 2004. *Savage girls and wild boys: A history of feral children*. London: Faber and Faber.

Nietzsche, F. W. 1986. *Human, all too human: A book for free spirits*. New York: Cambridge University Press. (Translated by R. J. Hollingdale; originally published 1878).

Nittrouer, S. 2001. Challenging the notion of innate phonetic boundaries. *Journal of the Acoustical Society of America*, 110, 1598–1605.

Noonan, J. P., Coop, G., Kudaravalli, S., Smith, D., Krause, J., Alessi, J., et al. 2006. Sequencing and analysis of Neanderthal genomic DNA. *Science*, 314, 1113–1121.

Núñez, R., and Sweetser, E. 2006. With the future behind them: Convergent evidence from Aymara language and gesture in the crosslinguistic comparison of spatial construals of time. *Cognitive Science*, 30, 401–450.

O'Brian. P. 1989. *The thirteen-gun salute*. London: Harper-Collins.

Okuda, J., Fujii, T., Ohtake, H., Tsukiura, T., Tanjii, K., Suzuki, K., et al. 2003. Thinking of the past and future: The roles of the frontal pole and the medial temporal lobes. *Neuroimage*, 19, 1369–1380.

Orwell, G. 1949. *Nineteen eighty-four, a novel*. London: Secker and Warburg.

Overskeid, G. 2007. Looking for Skinner and finding Freud. *American Psychologist*, 62, 590–595.

Papp, S. 2006. A relevance-theoretic account of the development and deficits of theory of mind in normally developing children and children with autism. *Theory and Psychology*, 16, 141–161.

Parker, E. S., Cahill, L., and McGaugh, J. L. 2006. A case of unusual autobiographical remembering. *Neurocase*, 12, 35–49.

Patterson, F. 1978. Conversations with a gorilla. *National Geographic*, 154, 438–465.

Patterson, N., Richter, D. J., Gnerre, S., Lander, E. S., and Reich, D. 2006. Genetic evidence for complex speciation of humans and chimpanzees. *Nature*, 441, 1103–1108.

Pavlov, I. P. 1927. *Conditioned reflexes*. Humphrey Milford: Oxford University Press.

Penn, D. C., Holyoak, K. J., and Povinelli, D. J. 2008. Darwin's mistake: Explaining the discontinuity between human and nonhuman minds. *Behavioral and Brain Sciences*, 31, 108–178.

Pepperberg, I. M. 1990. Some cognitive capacities of an African Grey parrot (Psittacus erithacus). *Advances in the Study of Behavior*, 19, 357–409.

———. 2000. *The Alex studies: Cognitive and communicative abilities of gray parrots*. Cambridge: Harvard University Press.

Petrides, M., Cadoret, G., and Mackey, S. 2005. Orofacial somatomotor responses in the macaque monkey homologue of Broca's area. *Nature*, 435, 1325–1328.

Pettit, P. 2002. When burial begins. *British Archaeology*, 66 (August).

Pettito, L. A., Zatorre, R. J., Gauna, K., Nikelski, E. J., Dostie, D., and Evans, A. C. 2000. Speech-like cerebral activity in profoundly deaf people processing signed languages: Implications for the neural basis of human language. *Proceedings of the National Academy of Sciences*, 97, 13961–13966.

Pietrandrea, P. 2002. Iconicity and arbitrariness in Italian Sign Language. *Sign Language Studies*, 2, 296–321.

Piaget, J. 1928. *Judgement and reasoning in the child*. London: Routledge and Kegan Paul.

Piattelli-Palmarini, M. (ed.) 1980. *The debate between Jean Piaget and Noam Chomsky*. Cambridge: Harvard University Press.

Pika, S., Liebal, K., and Tomasello, M. 2003. Gestural communication in young gorillas (*Gorilla gorilla*): Gestural repertoire, and use. *American Journal of Primatology*, 60, 95–111.

———. 2005. Gestural communication in subadult bonobos (*Pan paniscus*): Repertoire and use. *American Journal of Primatology*, 65, 39–61.

Pinker, S. 1994. *The language instinct*. New York: Morrow.

———. 1997. *How the mind works*. London: Penguin.

———. 2003. Language as an adaptation to the cognitive niche. In M. H. Christiansen and S. Kirby (eds.), *Language evolution* (pp. 16–37). Oxford: Oxford University Press.

———. 2007. *The stuff of thought: Language as a window into human nature*. London: Penguin.

Pinker, S., and Bloom, P. 1990. Natural language and natural selection. *Behavioral and Brain Sciences*, 13, 707–784.

Pinker, S., and Jackendoff, R. 2005. The faculty of language: What's special about it? *Cognition*, 95, 201–236.

Pinkham, A. E., Hopfinger, J. B., Pelphrey, K. A., Pivers, J., and Penn, D. L. 2008. Neural basis for impaired social cognition in schizophrenia and autism spectrum disorders. *Schizophrenia Research*, 99, 164–175.

Pizzuto, E., and Volterra, V. 2000. Iconicity and transparency in sign languages: A cross-linguistic cross-cultural view. In K. Emmorey and H. Lane (eds.), *The signs of language revisited: An anthology to honor Ursula Bellugi and Edward Klima* (pp. 261–286). Mahwah, NJ: Lawrence Erlbaum Associates.

Ploog, D. 2002. Is the neural basis of vocalisation different in non-human primates and *Homo sapiens*? In T. J. Crow (ed.), *The speciation of modern Homo Sapiens* (pp. 121–135). Oxford: Oxford University Press.

Poizner, H., Klima, E. S., and Bellugi, U. 1984. *What the hands reveal about the brain*. Cambridge: MIT Press.

Pollick, A. S., and de Waal, F. B. M. 2007. Apes gestures and language evolution. *Proceedings of the National Academy of Sciences*, 104, 8184–8189.

Povinelli, D. J. 2001. *Folk physics for apes*. New York: Oxford University Press.

Povinelli, D. J., and Bering, J. M. 2002. The mentality of apes revisited. *Current Directions in Psychological Science*, 11, 115–119.

Povinelli, D. J., Bering, J. M., and Giambrone, S. 2000. Toward a science of other minds: Escaping the argument by analogy. *Cognitive Science*, 24, 509–541.

Powell, A., Shennan, S., and Thomas, M. G. 2009. Late Pleistocene demography and the appearance of modern human behavior. *Science*, **324**, 1298–1301.

Premack, D. 1988. "Gavagai" or the future history of the animal language controversy. *Cognition*, **23**, 81–88.

———. 2007. Human and animal cognition: Continuity and discontinuity. *Proceedings of the National Academy of Sciences*, **104**, 13861–13867.

Premack, D., and Woodruff, G. 1978. Does the chimpanzee have a theory of mind? *Behavioral and Brain Sciences*, **4**, 515–526.

Provine, R. 2000. *Laughter: A scientific investigation*. London: Viking Penguin.

Pruetz, J. D., and Bertolani, P. 2007. Savanna chimpanzees, *Pan troglodytes verus*, hunt with tools. *Current Biology*, **17**, 412–417.

Ramachandran, V. S. 2000. Mirror neurons and imitation learning as the driving force behind "the great leap forward" in human evolution. *Edge*, no. 69, May 29.

Randi, J. 1982. *The magic of Uri Geller*. Amherst, NY: Prometheus Books.

Read, D. W. 2008. Working memory: A cognitive limit to non-human primate recursive thinking prior to hominid evolution. *Evolutionary Psychology*, **6**, 676–714.

Reichenbach, H. 1947. *Elements of symbolic logic*. New York: Macmillan.

Rhine, J. B. 1937. *New frontiers of the mind*. New York: Farrar and Rinehart.

Richmond, C. 2003. Obituary to David Horrobin. *British Medical Journal*, **326**, 885.

Rivas, E. 2005. Recent use of signs by chimpanzees (*Pan Troglodytes*) in interactions with humans. *Journal of Comparative Psychology*, **119**, 404–417.

Rizzolatti, G., and Arbib, M. A. 1998. Language within our grasp. *Trends in Neurosciences*, **21**, 188–194.

Rizzolatti, G., Camardi, R., Fogassi, L., Gentilucci, M., Luppino, G., and Matelli, M. 1988. Functional organization of inferior area 6 in the macaque monkey. II. Area F5 and the control of distal movements. *Experimental Brain Research*, **71**, 491–507.

Rizzolatti, G., Fogassi, L., and Gallese V. 2001. Neurophysiological mechanisms underlying the understanding and imitation of action. *Nature Reviews*, **2**, 661–670.

Rizzolatti, G., and Sinigaglia, C. 2006. *Mirrors in the brain: How our minds share actions and emotions*. Oxford: Oxford University Press.

Roberts, W. A., Feeney, M. C., MacPherson, K., Petter, M., McMillan, N., and Musolino, E. 2008. Episodic-like memory in rats: Is it based on when or how long ago? *Science*, **320**, 113–115.

Roediger, H. L., and McDermott, K. B. 1995. Creating false memories—remembering words not presented in lists. *Journal of Experimental Psychology: Learning Memory and Cognition,* **21**, 803–814.

Rousseau, J.-J. 1782. *Essai sur l'origine des langues*. Geneva.

Roy, A. C., and Arbib, M. A. 2005. The syntactic motor system. *Gesture*, **5**, 7–37.

Russell, B. A., Cerny, F. J., and Stathopoulos, E. T. 1998. Effects of varied vocal intensity on ventilation and energy expenditure in women and men. *Journal of Speech, Language, and Hearing Research,* **41,** 239–248.

Sacks, O. 1995. *An anthropologist on Mars.* New York: Vintage.

Saki (H. H. Munro) 1936. *The short stories of Saki.* New York: Viking.

Salmond, A. 1975. Mana makes the man: A look at Maori oratory and politics. In M. Bloch (ed.), *Political language and oratory in traditional society* (pp. 45–63). New York: Academic Press.

Saussure, F. de 1916. *Cours de linguistique générale,* ed. C. Bally and A. Sechehaye, with the collaboration of A. Riedlinger. Lausanne: Payot; translated as W. Baskin 1977, *Course in general linguistics.* Glasgow: Fontana/Collins.

Savage-Rumbaugh, S., Shanker, S. G., and Taylor, T. J. 1998. *Apes, language, and the human mind.* New York: Oxford University Press.

Savage-Rumbaugh, S., Wamba, K., Wamba, P., and Wamba, N. 2007. Welfare of apes in captive environments: Comments on, and by, a specific group of apes. *Journal of Applied Animal Science,* **10,** 7–19.

Schacter, D. L., Addis, D. R., and Buckner, R. L. 2007. Remembering the past to imagine the future: The prospective brain. *Nature Reviews Neuroscience,* **8,** 657–661.

Schaller, G. 1963. *The mountain gorilla.* Chicago: University of Chicago Press.

Schoenemann, P. T., Sheehan, M. J., and Glotzer, L. D. 2005. Prefrontal white matter volume is disproportionately larger in humans than in other primates. *Nature Neuroscience,* **8,** 242–252.

Schroeder, D. I., and Myers, R. M. 2008. Multiple transcription start sites for FOXP2 with varying cell specificities. *Gene,* **413,** 42–48.

Scoville, W. B., and Milner, B. 1957. Loss of recent memory after bilateral hippocampal lesions. *Journal of Neurology, Neurosurgery and Psychiatry,* **20,** 11–21.

Semaw, S., Renne, P., Harris, J. W. K., Feibel, C. S., Bernor, R. L., Fesseha, N., et al. 1997. 2.5-million-year-old stone tools from Gona, Ethiopia. *Nature,* **385,** 333–336.

Semendeferi, K., Damasio, H., and Frank, R. 1997. The evolution of the frontal lobes: A volumetric analysis based on three-dimensional reconstructions of magnetic resonance scans of human and ape brains. *Journal of Human Evolution,* **32,** 375–388.

Senju, A. Southgate, V., White, S., and Frith, U. 2009. Mindblind eyes: An absence of spontaneous theory of mind in Asperger Syndrome. *Science,* **325,** 883–885.

Senghas, A., Kita, S., and Ôzyürek, A. 2004. Children creating core properties of language: Evidence from an emerging sign language in Nicaragua. *Science,* **305,** 1780–1782.

Shaw, B. 1948. *Our theatres in the nineties.* Vol. 2. London: Constable and Co.

Sheldrake, R. 1999. *Dogs that know when their owners are coming home.* New York: Three Rivers Press.

Sheldrake, R., and Morgana, A. 2003. Testing a language-using parrot for telepathy. *Journal of Scientific Exploration*, 17, 601–615.

Shermer, M. 1997. *Why people believe weird things*. New York: Henry Holt.

Shi, R. S., Werker, J. F., and Cutler, A. 2006. Recognition and representation of function words in English-learning infants. *Infancy*, 10, 187–198.

Shintel, H., and Nusbaum, H. C. 2007. The sound of motion in spoken language: Visual information conveyed by acoustic properties of speech. *Cognition*, 105, 681–690.

Shintel, H., Nusbaum, H. C., and Okrent, A. 2006. Analog acoustic in speech. *Journal of Memory and Language*, 55, 167–177.

Sibley, C. G., and Ahlquist, J. E. 1984. The phylogeny of hominoid primates, as indicated by DNA-DNA hybridisation. *Journal of Molecular Evolution*, 20, 2–15.

Sims, M. 2003. *Adam's navel: A natural and cultural history of the human form*. London: Viking.

Skinner, B. F. 1957. *Verbal behaviour*. New York: Appleton-Century-Crofts.

———. 1962. *Walden two*. New York: Macmillan.

Slocombe, K. E., and Zuberbühler, K. 2007. Chimpanzees modify recruitment screams as a function of audience composition. *Proceedings of the National Academy of Sciences*, 104, 17228–17233.

Snow, C. P. 1979. *A coat of varnish*. London: Macmillan.

Snowdon, C. T. 2004. Social processes in the evolution of complex cognition and communication. In D. K. Oller and U. Griebel (eds.), *Evolution of communication systems* (pp. 132–150). Cambridge: MIT Press.

Sosis, R. 2004. The adaptive value of religious ritual. *American Scientist*, 92, 166–172.

Sousa, C., Biro, D., and Matsuzawa, T. 2009. Leaf-tool use for drinking water by wild chimpanzees (*Pan troglodytes*): Acquisition patters and handedness. *Animal Cognition*, 12 (Suppl. 1), 115.

Southgate, V., Senju, A., and Csibra, G. 2007. Action anticipation through attribution of false belief by 2-year-olds. *Psychological Science*, 18, 587–592.

Sperber, D., and Wilson, D. 1986. *Relevance: Communication and cognition*. Oxford: Blackwell.

———. 2002. Pragmatics, modularity and mind-reading. *Mind and Language*, 17, 3–23.

Squire, L. R. 1992. Declarative and nondeclarative memory—multiple brain systems supporting learning and memory. *Journal of Cognitive Neuroscience*, 4, 232–243.

Stedman, H. H., Kozyak, B. W., Nelson, A., Thesier, D. M., Su, L. T., Low, D. W., et al. 2004. Myosin gene mutation correlates with anatomical changes in the human lineage. *Nature*, 428, 415–418.

Stokoe, W. C. 2001. *Language in hand: Why sign came before speech*. Washington, DC: Gallaudet University Press.

Stokoe, W. C., Casterline, D. C., and Croneberg, C. G. 1965. *A dictionary of American Sign Language on linguistic principles*. Silver Spring, MD: Linstok Press.

Studdert-Kennedy, M. 1998. The particulate origins of language generativity: From syllable to gesture. In J. R Hurford, M. Studdert-Kennedy, and C. Knight (eds.), *Approaches to the evolution of language* (pp. 169–176). Cambridge: Cambridge University Press.

———. 2005. How did language go discrete? In M. Tallerman (ed.), *Language origins: Perspectives on evolution* (pp. 48–67). Oxford: Oxford University Press.

Suddendorf, T. 2006. Foresight and evolution of the human mind. *Science*, 312, 1006–1007.

———. 2010. Episodic memory versus episodic foresight: Similarities and differences. *WIRES Cognitive Science*, 1, 99–107.

Suddendorf, T., and Busby, J. 2003. Mental time travel in animals? *Trends in cognitive Sciences*, 7, 391–396.

Suddendorf, T., and Collier-Baker, E. 2009. The evolution of primate visual self-recognition: Evidence of absence in lesser apes. *Proceedings of the Royal Society B*, 276, 1671–1677.

Suddendorf, T., and Corballis, M. C. 1997. Mental time travel and the evolution of the human mind. *Genetic, Social, and General Psychology Monographs*, 123, 133–167.

———. 2007. The evolution of foresight: What is mental time travel, and is it unique to humans? *Behavioral and Brain Sciences*, 30, 299–351.

Suddendorf, T., Corballis, M. C., and Collier-Baker, E. 2009. How great is great ape foresight. *Animal Cognition*, 12, 751–754.

Sutton, D., Larson, C., and Lindeman, R. C. 1974. Neocortical and limbic lesion effects on primate phonation. *Brain Research*, 71, 61–75.

Sutton-Spence, R., and Boyes-Braem P. (eds.) 2001. *The hands are the head of the mouth: The mouth as articulator in sign language*. Hamburg: Signum-Verlag.

Szpunar, K. K., Watson, J. M., and McDermott, K. B. 2007. Neural substrates of envisioning the future. *Proceedings of the National Academy of Sciences*, 104, 642–647.

Tanner, J. E., and Byrne, R. W. 1996. Representation of action through iconic gesture in a captive lowland gorilla. *Current Anthropology*, 37, 162–173.

Tattersall, I. 2002. *The monkey in the mirror: Essays on the science of what makes us human*. New York: Harcourt.

Taylor, R., and Wiles A. 1995. Ring-theoretic properties of certain Hecke algebras. *Annals of Mathematics*, 141, 553–572.

Teleki, G. 1973. *The predatory behavior of wild chimpanzees*. Lewisburg, PA: Bucknell University Press.

Thieme, H. 1997. Lower Palaeolithic hunting spears from Germany. *Nature*, 385, 807–810.

Thompson, N. S. 1969. Individual identification and temporal patterning in the cawing of common crows. *Communications in Behavioral Biology*, 4, 29–33.

Thorpe, S.K.S., Holder, R. L., and Crompton, R. H. 2007. Origin of human bipedalism as an adaptation for locomotion on flexible branches. *Science*, **316**, 1328–1331.

Tomasello, M. 1999. *The cultural origins of human cognition*. Cambridge: Harvard University Press.

Tomasello, M. 2003. Introduction: Some surprises for psychologists. In M. Tomasello (ed.), *New psychology of language: Cognitive and functional approaches to language structure* (pp. 1–14). Mahwah, NJ: Lawrence Erlbaum.

———. 2008. *The origins of human communication*. Cambridge: MIT Press.

Tomaello, M., and Call, J. 1997. *Primate cognition*. New York: Oxford University Press.

Tomasello, M., Call, J., Warren, J., Frost, G. T., Carpenter, M., and Nagell, K. 1997. The ontogeny of chimpanzee gestural signals: A comparison across groups and generations. *Evolution of Communication*, **1**, 223–59.

Tomasello, M., Hare, B., and Agnetta, B. 1999. Chimpanzees, *Pan troglodytes*, follow gaze direction geometrically. *Animal Behaviour*, **58**, 769–777.

Tomasello, M., and Rakoczy, H. 2003. What makes human cognition unique? From individual to shared to collective intentionality. *Mind and Language*, **18**, 121–147.

Tooby, J., and DeVore, I. 1987. The reconstruction of hominid behavioral evolution through strategic modelling. In W. G. Kinzey (ed.), *The evolution of human behavior: Primate models* (pp. 183–237). New York: SUNY Press.

Toth, N., Schick, K. D., Savage-Rumbaugh, E. S., Sevcik, R. A., and Rumbaugh, D. M. 1993. Pan the tool-maker: Investigations into the stone tool-making and tool-using capabilities of a bonobo (*Pan paniscus*). *Journal of Archeological Science*, **20**, 81–91.

Treffert, D. A., and Christensen, D. D. 2006. Inside the mind of a savant. *Scientific American Mind*, **17**, 55–55.

Trivers, R. L. 1974. Parent-offspring conflict. *American Zoologist*, **14**, 249–264.

Tulving, E. 1983. *Elements of episodic memory*. New York: Oxford University Press.

———. 2001. Episodic memory and common sense: How far apart? *Philosophical Transactions of the Royal Society B: Biological Sciences*, **356**, 1505–1515.

———. 2002. Episodic memory: From mind to brain. *Annual Review of Psychology*, **53**, 1–25.

Tulving, E., Schacter, D. L., McLachlan, D. R., and Moscovitch, M. 1988. Priming of semantic autobiographical knowledge. *Brain and Cognition*, **8**, 3–20.

Uylings, H.B.M. (ed.) 1990. *The prefrontal cortex: Its structure, function, and pathology*. Amsterdam: Elsevier.

van Rijn, S., Swaab, H., and Aleman, A. 2008. Psychosis and autism as two developmental windows on a disordered social brain. *Behavioral and Brain Sciences*, **31**, 280–281.

Vargha-Khadem, F., Watkins, K. E., Alcock, K. J., Fletcher, P., and Passingham, R. 1995. Praxic and nonverbal cognitive deficits in a large family with a genetically transmitted speech and language disorder. *Proceedings of the National Academy of Sciences*, **92**, 930–933.

Volterra, V., Caselli, M. C., Capirci, O., and Pizzuto, E. 2005. Gesture and the emergence and development of language. In M. Tomasello and D. Slobin (eds.), *Beyond Nature-Nurture. Essays in Honor of Elizabeth Bates* (pp. 3–40). Mahwah, NJ: Lawrence Erlbaum.

Walenski, M., Mostofsky, S. H., Gidley-Larson, J. C., and Ullman, M. T. 2008. Brief report: Enhanced picture naming in autism. *Journal of Autism and Developmental Disorders*, **38**, 1395–1399.

Walsh, P. D., Abernethy, K. A., Bermejo, M., Beyersk, R., De Wachter, P., Akou, M. E., et al. 2003. Catastrophic ape decline in west equatorial Africa. *Nature*, **422**, 611–614.

Walter, R. C., Buffler, R. T., Bruggemann, J. H., Guillaume, M.M.M., Berhe, S. M., Negassi, B., et al. 2000. Early human occupation of the Red Sea coast of Eritrea during the last interglacial. *Nature*, **405**, 65–69.

Watkins, K. E., Strafella, A. P., and Paus, T. 2003. Seeing and hearing speech excites the motor system involved in speech production. *Neuropsychologia*, **41**, 989–994.

Watson, J. B. 1913. Psychology as the behaviorist views it. *Psychological Review*, **20**, 158–177.

Watson, J. B., and Rayner, R. 1920. Conditioned emotional reactions. *Journal of Experimental Psychology*, **3**, 1–14.

Wearing, D. 2005. *Forever today*. New York: Doubleday.

Wechkin, S., Masserman, J. H., and Terris, W. 1964. Shock to a conspecific as an aversive stimulus. *Psychonomic Science*, **1**, 47–48.

Weir, A. A. S., Chappell, J., and Kacelnik, A. 2002. Shaping of hooks in New Caledonian crows. *Science*, **297**, 981.

Westen, D. 1997. Toward a clinically and empirically sound theory of motivation. *International Journal of Psycho-Analysis*, **78**, 521–548.

Westergaard, G. C., Liv, C., Haynie M. K., and Suomi, S. J. 2000. A comparative study of aimed throwing by monkeys and humans. *Neuropsychologia*, **38**, 1511–1517.

Whiten, A., and Byrne, R. W. 1988. Tactical deception in primates. *Behavioral and Brain Sciences*, **11**, 233–244.

Whiten, A., Goodall, J., McGrew, W. C., Nishida, T., Reynolds, V., Sugiyama, Y., Tutin, C. E. G., Wrangham, R. W., and Boesch, C. 1999. Cultures in chimpanzees. *Nature*, **399**, 682–685.

Whiten, A., Horner, V., and de Waal, F. B. M. 2005. Conformity to cultural norms of tool use in chimpanzees. *Nature*, **437**, 737–740.

Whiten, A., McGuigan, N., Marshall-Pescini, S., and Hopper, L. M. 2009. Emulation, over-imitation and the scope of culture for child and chimpanzee. *Philosophical Transactions of the Royal Society B: Biological Sciences*, **364**, 2417–2428.

Whorf, B. L. 1956. *Language, thought and reality*. Cambridge: MIT Press.

Wiles, A. 1995. Modular elliptic curves and Fermat's last theorem. *Annals of Mathematics*, **141**, 443–451.

Wilson, D. S. 2002. *Darwin's cathedral: Evolution, religion, and the nature of society*. Chicago: University of Chicago Press.

Wittgenstein, L. 2005. *The big typescript, TS 213*. (C. G. Luckhardt and M. A. E. Aue eds. and trans.) Oxford: Basil Blackwood.

Wodehouse, P. G. 1922. *The clicking of Cuthbert*. London: Herbert Jenkins.

Wood, B., and Collard, M. 1999. The human genus. *Science*, **284**, 65–71.

Wrangham, R. 2009. *Catching fire: How cooking made us human*. New York: Basic Books.

Wundt, W. 1900. *Die Sprache*. 2 vols. Leipzig: Enghelman.

Wynn, C. D. L. 2004. *Do animals think?* Princeton, NJ: Princeton University Press.

Xua, J., Gannon, P. J., Emmorey, K., Smith, J. F., and Braun, A. R. 2009. Symbolic gestures and spoken language are processed by a common neural system. *Proceedings of the National Academy of Sciences*, **106**, 20664–20669.

Young, R. W. 2003. Evolution of the human hand: The role of throwing and clubbing. *Journal of Anatomy*, **202**, 165–174.

Zeshan, U. 2002. Sign language in Turkey: The story of a hidden language. *Turkic Languages*, **6**, 229–74.

Zipf, G. K. 1949. *Human behavior and the principle of least-effort*. New York: Addison-Wesley.

Index

The Recursive Mind

The Origins of Human Language, Thought, and Civilization

Michael C. Corballis

"Corballis offers a novel synthesis of language, mental time travel, and theory of mind within an evolutionary perspective. *The Recursive Mind* is very well written for a general readership, but with lots of targeted references for experts."
—Michael A. Arbib, coauthor of *The Construction of Reality*

"This is a wonderful book by an expert writer. Corballis tracks the importance of recursion in the context of language, theory of mind, and mental time travel, and concludes that its emergence explains much about how we became human. He proposes a novel answer to an enduring mystery. This book is a significant achievement."
—Thomas Suddendorf, University of Queensland

PRINCETON

press.princeton.edu

ISBN 978-0-691-14547-1

90000

9 780691 145471